Space Law and Policy

Series Editor

Maria A. Pozza, Gravity Lawyers, Christchurch, New Zealand

Space activities across the Asia-Pacific, specifically in Australia and New Zealand, are increasing and diversifying. Over the last decade, countries in this region have seen significant growth within their national space industries and economies which in turn benefit their wider societies due to the downstream application and use of, space-related technologies, products, and services. This phenomena presents not only a great opportunity for these countries, but also represents a situation necessitating considered and fit-for-purpose space law and policy. Robust yet flexible regulatory regimes are needed if countries in the Asia-Pacific, such as Australia and New Zealand, are to benefit from the strategic and commercial advantages associated with a presence in outer space and functioning domestic space sector. These regulatory regimes must respond to, and address, numerous challenges ranging from adherence to international legal obligations to issues of geo-political space security and military concerns through to questions about how best to regulate and support commercial space endeavours. These myriad factors will have an impact upon how countries develop, implement, and refine their space laws and policies. This series offers a unique perspective from this region and intends to act as a source of knowledge and guidance for space actors across government, the private sector and academia in the pursuit and furtherance of their respective space activities and operations. The contributors to this series are all experts in their field who share a common interest in strengthening space governance via the promotion and development of space law and policy.

Hamid Jahankhani · Stefan Kendzierskyj ·
Sina Pournouri · Maria A. Pozza
Editors

Space Law Principles and Sustainable Measures

Springer

Editors
Hamid Jahankhani
Northumbria University London
London, UK

Stefan Kendzierskyj
Cyfortis
Worcester Park, UK

Sina Pournouri
Sheffield Hallam University
Sheffield, UK

Maria A. Pozza
Gravity Lawyers
Christchurch, New Zealand

ISSN 2662-9054 ISSN 2662-9062 (electronic)
Space Law and Policy
ISBN 978-3-031-64044-5 ISBN 978-3-031-64045-2 (eBook)
https://doi.org/10.1007/978-3-031-64045-2

This Springer imprint is published by the registered company Springer Nature Switzerland AG
The registered company address is: Gewerbestrasse 11, 6330 Cham, Switzerland

Contents

The Principle of 'Due Regard' in the Light of the Artemis Accords for a Sustainable Lunar Exploration

Anne-Sophie Martin

Abstract The chapter aims to analyse the concept of 'due regard' as introduced in Article IX of the 1967 Outer Space Treaty and how it is illustrated in the Artemis Accords. The legal definition and the notion of "due regard for the corresponding interests of all other States" are still under debate. In practice, entities conducting space missions hope that no one will interfere with their activities, whether accidentally or intentionally. The chapter focuses on what constitutes the concept of 'due regard', for instance through the protection of cultural heritage, and the creation of safety zone. Those elements represent a practical application of the concept and are reflected in the Artemis Accords. In a certain way, soft law instruments, as the Artemis Accords, might lead to the harmonisation of the Law of Outer Space and might contribute to the application of core concepts present in the Outer Space Treaty, such as the principle of 'due regard'. The development of non-binding instruments, containing principles first introduced in hard law, might lead to the acceptance and application by States and private entities of the main principles of space law, thus contributing to a sustainable lunar exploration and beyond.

Keywords Due regard · International law · Space law · Information sharing · Cultural heritage · Safety zone

A.-S. Martin (✉)
Department of Political Sciences, Sapienza University of Rome, Rome, Italy
e-mail: martin.annesophie@yahoo.fr

Institute for International Legal Studies (ISGI-CNR), Rome, Italy

Institute on Space Law and Ethics-For All Moonkind, Oxford, USA

H. Jahankhani et al. (eds.), *Space Law Principles and Sustainable Measures*, Space Law and Policy, https://doi.org/10.1007/978-3-031-64045-2_1

1 Introduction

A fast-growing number of countries,[1] space agencies and private actors[2] are interesting in joining space explorations activities, including mining operations, human settlement, missions *in situ* on the Moon and in orbit, notably in the context of the Artemis programme.[3]

With the Artemis programme, the United States aims of returning astronauts to the Moon[4] and establishing a permanent base there. A space station, the *Lunar Orbital Platform-Gateway* (LOP-G), will be placed in orbit around the Moon. It will serve as a relay between the Earth and the Moon, and as a fuel supply point. The construction of a lunar base represents the first step in space exploration, with the objective to continue the mission and then move on to Mars. The Artemis programme lays the foundations for sustainable manned and robotic exploration of our natural satellite and includes the development of several spacecrafts: the heavy-lift *Space Launch System* (SLS), the *Orion* spacecraft, the *Human Landing System* (HLS) designed to take humans to the Moon, and robotic missions to carry out reconnaissance and complementary scientific studies.

In February 2024, a new milestone has been achieved in lunar exploration. The US space company, Intuitive Machines, landed on the Moon with the Odysseus module. This represents the first privately developed spacecraft to touch down on the Moon and the first US soft landing since 1972.[5]

China has also an ambitious space exploration programme after the success of the landing of Chang'e-4 spacecraft on the far side of the Moon in 2019, and its Earth-orbiting space station, Tiangong, which was completed in November 2022. In 2020, China's National Space Administration reiterated the country's ambition to establish an international scientific station on the Moon (ILRS) from 2036 which is scheduled to be deployed at the South Pole and to exploit its resources.[6] Russia joined the project in 2021.[7]

---

[1] United-States, Russia, China, Japan, Israel and India.

[2] Rebecca Boyle, 'A New Private Moon Race Kicks Off Soon' (Scientific American, 2022), www.scientificamerican.com/article/a-new-private-moon-race-kicks-off-soon/, accessed 21 February 2024. e.g. Ispace, Astrobotic.

[3] NASA Artemis Program, www.nasa.gov/specials/artemis/, accessed 21 February 2024.

[4] US Department of State, 'International Cooperation in NASA's Artemis I Program' (16 November 2022), www.state.gov/international-cooperation-in-nasas-artemis-i-program/, accessed 21 February 2024.

[5] Jeff Foust, 'Intuitive Machines Lands on the Moon' (SpaceNews, 22 February 2024), https://spacenews.com/intuitive-machines-lands-on-the-moon/, accessed 23 February 2024.

[6] International Lunar Research Station (ILRS), 'Deep Space Exploration Laboratory of CNSA' (CNSA, May 2023), https://www.unoosa.org/documents/pdf/copuos/2023/TPs/ILRS_presentation20230529_.pdf, accessed 21 February 2024.

[7] Scott Neuman, 'China, Russia Announce Plan to Build Moon Research Stations' (NRP, 10 March 2021), www.npr.org/2021/03/10/975579975/china-russia-announce-plan-to-build-moon-research-station, accessed 21 February 2024.

With this in mind, our natural satellite could soon become a crowded place with human and robotic activities. In fact, the development of new space activities on the Moon, particularly those associated with mining, poses organisational challenges given the expected geographical clustering of resources.[8]

Over the next few years,[9] numerous missions are being planned on the Moon.[10] A number of these missions include permanent lunar bases and intend to explore, exploit and use the Moon's natural resources in situ, including water[11] and helium-3,[12] or return them to Earth.[13] Initially, space resources could be used to support and develop projects on lunar surface, but in the longer term, the Moon and its resources will constitute key element for deep space missions. In fact, lunar and cis-lunar activities represent a first stage towards Mars, asteroids, comets and other celestial bodies.

Ongoing research reveals that there are relatively small zones that contain useful materials[14] for sustaining life on the Moon. This concentration of resources could lead to situation of tensions between actors operating on its surface, as many of the missions are probably exploring and prospecting the same areas of the Moon based on where resources are located. This could have a negative impact on the safety and sustainability of lunar activities.

In addition, technical issues are potentially arising due to the absence of atmosphere on the Moon, implying that dust and rock particles can have significant negative effects on activities.[15] Indeed, dust is easily kick up when spacecraft land or

[8] Harvard-Smithsonian Center for Astrophysics, 'Growing Interest in Moon Resources Could Cause Tension, Scientists Find' (PhysOrg, 23 November 2020), https://phys.org/news/2020-11-moon-resources-tension-scientists.html, accessed 21 February 2024.

[9] Michelle Hanlon, 'Lunar Mining and Moon Land Claims Fall into a Gray Area of International Law, But Negotiations are Underway to Avoid Conflict and Damage to Spacecraft', (The Conversation, 23 August 2022), https://theconversation.com/lunar-mining-and-moon-land-claims-fall-into-a-gray-area-of-international-law-but-negotiations-are-underway-to-avoid-conflict-and-damage-to-spacecraft-188426, accessed 21 February 2024.

[10] NSR, 'Developing Moon Market Propelled by 2050 + Missions and $105 Billion in Revenue' (10 April 2023), www.nsr.com/nsr-developing-moon-market-propelled-by-250-missions-and-105-billion-in-revenue-through-decade/, accessed 21 February 2024.

[11] Leonard Davic, 'How Water on the Moon Could Fuel Space Exploration' (Space, 29 March 2012), www.space.com/15094-moon-water-ice-space-fuel.html, accessed 21 February 2024.

[12] European Space Agency, 'Helium-3 Mining on the Lunar Surface', www.esa.int/Enabling_Support/Preparing_for_the_Future/Space_for_Earth/Energy/Helium-3_mining_on_the_lunar_surface, accessed 21 February 2024.

[13] Angel Abbid-Madrid, 'Space Resource Utilization' (Oxford Research Encyclopedias, 28 June 2021), https://doi.org/10.1093/acrefore/9780190647926.013.13.

[14] Harvard-Smithsonian Center for Astrophysics, 'Growing Interest in Moon Resources Could Cause Tension, Scientists Find' (PhysOrg, 23 November 2020), https://phys.org/news/2020-11-moon-resources-tension-scientists.html, accessed 21 February 2024.

[15] NASA, Apollo 17 Technical Crew Debriefing (NASA, 4 January 1973), https://an.rsl.wustl.edu/apollo/data/A17/resources/a17-techdebrief.pdf, accessed 21 February 2024; NASA Technical Reports Server, Results of Examination of the Returned Surveyor 3 Samples for Particulate Impacts (NASA, 1972), www.lpi.usra.edu/lunar/strategies/Cour-PalaisEtAl_NASA-SP-284E_Surveyor_3_Particulate_Impacts.pdf, accessed 21 February 2024; Michelle Hanlon and Bailey Cunningham.

roll on the lunar surface[16] and it can have a severe impact on machinery and human activities,[17] and might also deteriorate historic lunar sites.

Therefore, against this background, where an ever increasing number of activities are foreseen on the Moon and keeping in mind the risks of damage[18] and harmful interferences,[19] it is necessary to establish rules to govern lunar activities in order to ensure that activities conducted by States and private entities are pursued in a safe, secure and sustainable manner.[20] The 2019 Long-Term Sustainability Guidelines define the sustainability of outer space activities as the ability to maintain the conduct of space activities indefinitely into the future in a manner that realizes the objectives of equitable access to the benefits of the exploration and use of outer space for peaceful purposes, in order to meet the needs of present generations while preserving the outer space environment for future generations.[21]

In particular, one of a core principle that need to be applied is the principle of "due regard to the corresponding interests of all other States" as mentioned in Article IX of the 1967 *Treaty on Principles Governing the Activities of States in the Exploration*

'The Legal Imperative to Mitigate the Plume Effect: An 'Aggravation and Frustration' That Imperils Our History and Our Future' (2019) 43 Journal of Space Law, 309.

[16] Center for Lunar and Asteroid Surface Science, University of Central Florida, 'The Science of Plume Effects', https://sciences.ucf.edu/class/landing-team/the-science-of-plume-effects/, accessed 21 February 2024.

[17] NASA, Apollo 17 Technical Crew Debriefing (NASA, 4 January 1973), https://an.rsl.wustl.edu/apollo/data/A17/resources/a17-techdebrief.pdf, accessed 21 February 2024; NASA Technical Reports Server, Results of Examination of the Returned Surveyor 3 Samples for Particulate Impacts (NASA, 1972), www.lpi.usra.edu/lunar/strategies/Cour-PalaisEtAl_NASA-SP-284E_Surveyor_3_Particulate_Impacts.pdf, accessed 21 February 2024; Michelle Hanlon and Bailey Cunningham. 'The Legal Imperative to Mitigate the Plume Effect: An 'Aggravation and Frustration' That Imperils Our History and Our Future' (2019) 43 Journal of Space Law, 309.

[18] Convention on International Liability for Damage Caused by Space Objects, Article I (a), The term "damage" means loss of life, personal injury or other impairment of health; or loss of or damage to property of States or of persons, natural or juridical, or property of international intergovernmental organizations.

[19] "Radio interference" is defined by provision No. 1.166 of the ITU Radio Regulations as "the effect of unwanted energy due to one or a combination of emissions, radiations, or inductions upon reception in a radiocommunication system, manifested by any performance degradation, misinterpretation or loss of information which could be extracted in the absence of such unwanted energy". "Harmful interference" is defined in both No. 1.169 of the RR and in No.1003 of the ITU Constitution, as "interference which endangers the functioning of a radionavigation service or of other safety services or seriously degrades, obstructs, or repeatedly interrupts a radiocommunication service operating in accordance with Radio Regulations". Harmful interference, the blocking, jamming or degradation of service can take place for a wide range of reasons, which can be either accidental or intentional.

[20] Jessy Kate Schinger, 'Imagining Safety Zones: Implications and Open Questions' (The Space Review, 8 June 2020), www.thespacereview.com/article/3962/1, accessed 21 February 2024.

[21] *See* UN Doc. A/74/20 and UN Doc. A/AC.105/C.1/L.366.

and Use of Outer Space, including the Moon and Other Celestial Bodies (Outer Space Treaty—OST).[22]

In this race to the Moon, space exploration and lunar activities cannot become a far west, without rules, it has to be conducting "for the benefit and in the interest of all countries, irrespective of their degree of economic or scientific development […]".[23]

These new projects shed light on the fact that the Moon environment could be threatened, jeopardising the Earth-Moon ecosystem which is primordial for life on Earth and biosphere. Given the rise of missions and actors involved, it is necessary to consider the adoption of lunar exploration rules, illustrated notably by the Artemis Accords, and in the 'operationalisation' of the principle of 'due regard'.

The Artemis Accords,[24] signed in October 2020, set out the legal framework that will apply to activities that should be carried out in situ on the Moon, but also in orbit. They represent in a way the normative aspect of the Artemis programme. These rules could be extended to other deep-space exploration missions on Mars, comets and asteroids. As of June 2024, 43 States have signed the Artemis Accords.[25]

In this respect, challenges are to guarantee the sustainability of lunar activities and to ensure that operations are conducted in a safe and secure manner without causing interferences with missions of other actors involved in lunar exploration.

The chapter aims to examine the principle of 'due regard' under international law and space law, its components, and how the Artemis Accords represent a practical application of the principle considering the sharing of information, the protection of cultural heritage and the concept of safety zone.

2 Some Thoughts on the Meaning of 'Due Regard' Under International Law

The first international Convention making reference to the principle of 'due regard' is the 1944 Convention on International Civil Aviation,[26] known as the Chicago Convention, which granted the International Civil Aviation Organization (ICAO) the competences to establish principles and procedures for the safety of air navigation.

---

[22] Treaty on Principles Governing the Activities of States in the Exploration and Use of Outer Space, Including the Moon and Other Celestial Bodies (Outer Space Treaty), 27 January 1967, 610 UNTS 205, in force 10 October 1967.

[23] Article I of the OST.

[24] The Artemis Accords, Principles for cooperation in the civil exploration and use of the Moon, Mars, Comets and Asteroids for peaceful purposes (NASA, 13 October 2020), https://www.nasa.gov/wp-content/uploads/2022/11/Artemis-Accords-signed-13Oct2020.pdf, accessed 21 February 2024.

[25] Robert Lea, 'Artemis Accords: What are They & Which Countries are Involved?' (Space, 16 February 2024), https://www.space.com/artemis-accords-explained, accessed 21 February 2024.

[26] Convention on International Civil Aviation, 7 December 1944, 15 UNTS 295, in force 4 April 1947.

The Article 3 of the Convention deals with civil and State aircrafts. It provides in para. (d) that "*the contracting States undertake, when issuing regulation for their state aircraft, that they will have due regard for the safety of navigation of civil aircraft*". ICAO rules are applicable only for civil aircraft. However, State aircrafts (identified by ICAO as military or policy services in Article 3 (b)) must have 'due regard' for the safety of navigation of civil aircraft.[27] In other words, State aircraft must act with due diligence for the 'safety of navigation of civil aircraft' by (i) operating like other civil aircraft; (ii) operating inside reserved or restricted airspace; or (iii) using special procedures established the State.[28]

The 'due regard' principle is the main treaty obligation imposed upon States for the regulation of the flight of military aircraft applicable during times of peace and armed conflict.[29] States parties have therefore accepted the obligation to include 'due regard' provision in their national legal regime when they establish regulation affecting the flight of State aircraft.[30] Article 3(d) creates an obligation on States to regulate their aircraft in order to ensure that the aircraft exercise appropriate attention, as well as, heed and care for the safety of civil aircraft, in particular in avoiding obstruction to the course of and collisions with civil aircraft.[31]

The concept of 'due regard' can also be found in the 1982 United Nations Convention on the Law of the Sea (UNCLOS).[32] Article 87 of the UNCLOS stresses that "*The high seas are open to all States [...]*". Para. 2 of the Article precises that "*These freedoms shall be exercised by all States with due regard for the interests of other States in their exercises of the freedom of the high seas [...]*". The high seas and outer space share the same legal nature, and in both areas there is a necessity to conduct activities with due regard, considering the activities carry out by other actors.

At international level, court and tribunal have examined the definition and the criteria for applying the principle. Indeed, the International Court of Justice (ICJ), in the 1974 Fisheries Jurisdiction case[33] refers to the obligation to pay 'due regard' to the interests of other States in the conservation and equitable exploitation of these resources and recalls that:

> It is one of the advances in maritime international law, resulting from the intensification of fishing, that the former laissez faire treatment of the living resources of the sea in the high seas has been replaced by a recognition of a duty to have due regard to the rights of other States and the needs of conservation for the benefit of all. Consequently, both Parties have the

[27] ICAO, MIDANPIRG/18&RASG-MID/8_WP/46 (9 February 2021), Due Regard Operation, Awareness and Associated Regulations.

[28] State Aircraft and 'Due Regard' Operations, 57th Annual Conference, Accra, Ghana, 19–23 march 2018, WP n°92, https://ifatca.wiki/kb/wp-2018-92/, accessed 21 February 2024.

[29] Michel Bourbonniere and Louis Haeck, 'Military Aircraft and International Law: Chicago Opus 3' (2001) 66 Journal of Air Law and Commerce 3, 913.

[30] Ibid., 915.

[31] Ibid., 916.

[32] United Nations Convention on the Law of the Sea, 10 December 1982, 1833 UNTS 396, in force 16 November 1994.

[33] ICJ, Fisheries Jurisdiction Case (*United Kingdom of Great Britain and Northern Ireland v. Iceland*), Merits, Judgment, Reports 1974, p. 3.

> obligation to keep under review the fishery resources in the disputed waters and to examine together, in the light of scientific and other available information, the measures required for the conservation and development, and equitable exploitation, of those resources [...][34]

In the 2015 Chagos Marine Protected Areas Arbitration,[35] a tribunal of the Permanent Court of Arbitration found that the principle of 'due regard' or 'reasonable regard' is composed of four elements[36]: (i) the nature of the rights held by whom is being harmed, (ii) the extent of the anticipated impairment to whom is being harmed, (iii) the nature and significance of the activities of the State causing the harm and, (iv) other possible approaches. The principle does not prevent the activity of the first State from causing damage, but indicates when the second State must be consulted on the extent of the damage.[37]

The application of the notion of 'due regard' implies that a State, in exercising a right, is required to take into account the existence of conflicting rights and interests of other States and to balance their respective significance. The obligations of 'due regard' share many characteristics between the different international law regimes. Their purpose is to ensure conciliation between conflicting concurrent or overlapping elements, particularly rights belonging to different States. Showing 'due regard' entails a duty to cooperate and consultation to strike the most appropriate balance between the divergent rights or obligations at stake, including areas beyond national jurisdiction such as the high seas and outer space.

3 The Principle of 'Due Regard' Under Article IX of the Outer Space Treaty

Similar to Article 87(2) of UNCLOS, Article IX of the OST provides that "*State Parties shall conduct all their activities in outer space with due regard to the corresponding interests of all other States*". This means that States, as well as private entities authorized by the appropriate State, should conduct their space activities with a certain standard of care, attention or observance to others.[38] In this respect, Article IX restrains the freedom of exploration and use of outer space enshrined in Article I para.2 of the OST. Outer space must be explored and used with due diligence by taking into consideration the interests and rights of other States. States have an obligation to ensure that the exercise of their rights and freedoms in outer space does

[34] ICJ, Fisheries Jurisdiction Case, 72.

[35] PCA, Chagos Marine Protected Area Arbitration (*Mauritius v. United Kingdom*), 18 March 2015.

[36] *See* paras. 471 and 519.

[37] Interview with Dan Hawk, Due Regard for the earth-Moon Ecosystem, PressBooks, https://pressbooks.pub/indigenizingclimateaction/chapter/due-regard-for-the-earth-moon-ecosystem/, accessed 21 February 2024.

[38] Sergio Marchisio, 'Article IX', in Stephan Hobe, Bernhard Schmidt-Tedd, Kai-Uwe Schrogl and Gérardine Meishan Goh (eds), *CoCoSL*, Vol. 1 (Carl Heymanns Verlag 2009) 175.

not interfere with, or jeopardize the safety of space missions of other States.[39] From this point of view, the freedom to use outer space entails that a State has the right to undertake activities that do not hinder the operations of other States.

Moreover, the terms 'corresponding interests', mentioned in Article IX, imply that State should carry out space operations in accordance with its own interest, as well as in the interests and rights of other States conducting space activities.

Article IX also defines the notion of 'international consultations' between parties in two cases: first, when a State Party "*has reason to believe that an activity or experiment planned by it or its nationals in outer space, including the moon and other celestial bodies, would cause potentially harmful interference with activities of other States Parties in the peaceful exploration and use of outer space*" and, second, when a State Party "*has reason to believe that an activity or experiment planned by another State Party in outer space, including the moon and other celestial bodies, would cause potentially harmful interference with activities in the peaceful exploration and use of outer space*". Consequently, consultations must be held before the activities in question actually take place, as well as at any stage, even during missions. Furthermore, State has to demonstrate, beyond any reasonable doubt, that all possible efforts have been deployed to preclude a harmful act from being committed.[40]

The application of the principle of 'due regard' might substantially mitigate threats arising from State behaviours in conducting their space missions.[41] Since 'due regard' is specifically addressed as a legal obligation under Article IX, a breach of the principle in the conduct of space activity could be seen as a violation of the Outer Space Treaty that could trigger the international responsibility of the State that caused the infringement. However, the concept lacks of meaning, contents and criteria to concretely be applied by States and private actors. It is therefore necessary to specify the measures required for the effective application of the 'due regard' principle in space activities.

By applying the principle of 'due regard' to space activities, it is first necessary to consider outer space as a unique environment, located beyond national jurisdiction. Then, the 'due regard' principle represents a duty of care and self-restrain in the exercise of the rights granted to States, implying consultation and international cooperation in case of risks of interference in the conduct of space operations. In other terms, it consists of balancing different States or private entities' interests and rights, while making sure that an actor, involved in lunar programme, exercises its rights without causing harmful interference with activities of other actors.[42] Those elements play also a significant role for ensuring that space activities are conducted

[39] Ibid.

[40] Ibid.

[41] Setsuko Aoki, Presentation "How Can the Principle of Due Regard Be Applied to Address Threats Arising from State Behaviours with Respect to Outer Space?", UNODA, 9 May 2022, https://documents.unoda.org/wp-content/uploads/2022/05/Panel1-Setsuko-Aoki_09052022-final.pdf, accessed 21 February 2023.

[42] Ibid.

in a safe, secure and sustainable way as it is needed to consider the rights and interests of other entities in carrying out space missions.

Furthermore, Article IX must be read in conjunction with Article XI of the OST which provides that "*States parties conducting activities in outer space agree to inform the Secretary-General of the United Nations as well as the public and the international scientific community, to the greatest extent feasible and practicable, of the nature, conduct, locations and results of such activities*". Articles IX and XI work together for 'due regard' principle, particularly for what concerns the sharing of information and the availability of scientific data.

The Artemis Accords represent a non-legally binding instrument introducing notions that contribute to a practical application of the 'due regard' principle to lunar activities.

4 Practical Application of the Principle of 'Due Regard' Within the Artemis Accords

The Artemis Accords contain four sections that are particularly relevant in the 'operationalization' of the principle of 'due regard'.

4.1 Transparency and Information Sharing

The first element to consider is the concept of transparency in conducting space missions and the sharing of information. The Accords focuses on these topics in Sect. 4 'Transparency' which specifies that

> the Signatories plan to share scientific information resulting from their activities with the public and the international scientific community on a good-faith basis and consistent with Article XI of the Outer Space Treaty;

and Sect. 8 'Release of scientific data'. It provides that

> the Signatories retain the right to communicate and release information to the public regarding their own activities. The Signatories intend to coordinate with each other in advance regarding the public release of information that relates to the other Signatories' activities [...]. The Signatories are committed to the open sharing of scientific data. The Signatories plan to make the scientific results obtained from cooperative activities under these Accords available to the public and the international scientific community, as appropriate, in a timely manner.

Information sharing represents a practical application of the principle of 'due regard' and contributes to the promotion of international cooperation and coordination among States. The system of notification and information allow space operators to take appropriate measures to protect their interests. With this in mind, States need to begin initial step towards disseminating information about the nature, conduct, location and results (data parameters to share) of their lunar mission plans. Consequently, an initial set of mission data which, once made public, could support space operators in the overall process of mitigating and avoiding interference . Lastly, it will be necessary to coordinate this system with non Signatories party to the Artemis Accords.

4.2 The Preservation of Historic Artifacts on Celestial Bodies

A further section to be noted in the Accords is Sect. 9 "Preserving Outer Space Heritage" which emphasises that

> The Signatories intend to preserve outer space heritage, which they consider to comprise historically significant human or robotic landing sites, artifacts, spacecraft and other evidence of activity on celestial bodies [...]. The Signatories intend [...] to contribute to multilateral efforts to further develop international practices and rules applicable to preserving outer space heritage.

With the development of lunar exploration activities over the next few years, the international community is realising the need to protect cultural sites on the Moon and other celestial bodies.[43] Apart from the landing sites of the Apollo missions,[44] it is reckoned that there are over 110 potential human heritage sites on the Moon.[45] Therefore, the question emerges of the need to safeguard, protect and preserve such cultural and heritage sites. The aim is to avoid that a lander, a robot, an astronaut, or an installation alter cultural heritage. Indeed, the historic sites and artifacts of past human missions of exploration deserve some measure of conservation.[46] This provision and the need to protect lunar artifacts are based on Article IX of the OST, which prohibits 'harmful contamination' and 'harmful interference', as well as the application of the principle of 'due regard'.[47]

---

[43] *See* For All Moonkind website, a US not-for-profit organization aimed at developing norms to protect lunar heritage sites, including with a possible convention, or recognition by UNESCO, www.forallmoonkind.org/, accessed 21 February 2024.

[44] *See* For All Moonkind, Crewed Landing Sites, www.forallmoonkind.org/moonkind-mission/apollo-landing-sites/, accessed 21 February 2024; *see also* NASA History Program Office, Catalogue of Manmade Material on the Moon (NASA, July 2012), https://history.nasa.gov/FINAL%20Catalogue%20of%20Manmade%20Material%20on%20the%20Moon.pdf, accessed 21 February 2024; *see* For All Moonkind, Moon Registry, https://moonregistry.forallmoonkind.org/, accessed 21 February 2024.

[45] *See* For All Moonkind, Other Landing Sites, www.forallmoonkind.org/moonkind-mission/other-landing-sites/, accessed 21 February 2024.

[46] Anne-Sophie Martin, 'The 'Outstanding Universal Value' Concept of the UNESCO World Heritage Convention: Food for Thought to Preserve Lunar Artifacts', in Annette Froehlich (ed.), *Protection of Cultural Heritage Sites on the Moon* (Springer, 2020) 53–64.

[47] Christopher Johnson, 'The Space Law Context of the Artemis Accords', Part. 2 (SpaceWatch, 28 May 2020), https://spacewatch.global/2020/05/spacewatchgl-feature-the-space-law-context-of-the-artemis-accords-part-2/, accessed 21 February 2024.

At international level, the legal Subcommittee of the United Nations Committee on the Peaceful Uses of Outer Space (UNCOPUOS) acknowledged, in March 2023, the necessity to recognize and to encourage the adoption of rules to protect sites that represent the history of space exploration.[48]

At national level, the United States, a precursor in this regard, is committed to elaborating standards for the protection of historic sites on celestial bodies. For instance, NASA adopted in 2011 a set of recommendations to protect Apollo and other U.S. objects on the lunar surface that are of historical and scientific value.[49] In addition, the United States recently passed the "One Small Step to Protect Human Heritage in Space Act" which requires NASA to ensure that its contractors do not damage lunar heritage sites.[50] However, these proposals are not binding on any entity or nation unless it contracts with NASA.

This is just the first step, because in addition to these major national measures, the question remains as to whether an international framework can be put in place to provide protection to lunar heritage sites[51] as lunar artifacts encompass US, Soviet, Indian, Chinese, Japanese, Isreali spacecrafts, so it would be a recognition in the interest to the international community as a whole.

4.3 *The Creation of Safety Zone*

The creation of safety zone can be seen as "operationalizing" and putting in practice the principles of due regard and international cooperation, two core elements of Article IX of the OST.[52] Indeed, safety zone will allow conducting space operations in a safe and secure manner, taking into account the interest of all other States, mainly

---

[48] Statement by the Delegation For All Moonkind, 28 March 2023, https://www.unoosa.org/documents/pdf/copuos/lsc/2023/Statements/28_AM/12_For_All_Moonkind_28_Mar_AM_.pdf, accessed 21 February 2024; *see also* UN Doc. A/AC.105/C.2/2023/CRP.35, 20 March 2023.

[49] NASA's Recommendations to Space-Faring Entities: How to Protect and Preserve the Historic and Scientific Value of U.S. Government Lunar Artifacts (NASA, 20 July 2011), www.nasa.gov/pdf/617743main_NASA-USG_LUNAR_HISTORIC_SITES_RevA-508.pdf, accessed 21 February 2024.

[50] S. 1694—One Small Step to Protect Human Heritage in Space Act, Public Law No: 116–275 (31 December 2020), https://www.congress.gov/116/plaws/publ275/PLAW-116publ275.pdf, accessed 21 February 2024.

[51] Anne-Sophie Martin, 'Vers un nouveau cadre juridique sur la protection des sites et biens présents sur les corps célestes?', in *Liber Amicorum Sergio Marchisio*, Vol.II (Ed. Scientifica Napoli, 2022) 1121–1138.

[52] Lucas Mallowan, Lucien Rapp and Maria Topka, 'Reinventing Treaty Compliant "Safety Zones" in the Context of Space Sustainability' (2021) 8 Journal of Space Safety Engineering 2, 155–166, https://doi.org/10.1016/j.jsse.2021.05.001; Michelle L.D. Hanlon, 'Due Regard and Safety Zones: Understanding the Commercial Implications of Recent Policy and Legislation' (Presentation NASA, 13 January 2021), www.nasa.gov/sites/default/files/atoms/files/space_portal_michelle_hanlon.pdf, accessed 21 February 2024; Jessy Kate Schingler, 'Imagining Safety Zones: Implications and open questions' (The Space Review, 8 June 2020), https://www.thespacereview.com/article/3962/1, accessed 21 February 2024.

by avoiding the risk of harmful interference between activities of actors involved. The establishment of these areas would necessitate coordination between all operators. These consultations would enable to determine and to agree on the characteristics of a safety zone.

Section 11 'Deconfliction of Space Activities' of the Artemis Accords specifically mentions Article IX of the OST and the notion of due regard, harmful interference, as well as the mechanism of consultation in the creation of safety zone. It outlines how safety zones could be implemented to limit harmful interference between multiple space operations and facilitate the exploitation of the Moon by various nations. In particular, it stresses that

> 7. In order to implement their obligations under the Outer Space Treaty, the Signatories intend to provide notification of their activities and commit to coordinating with any relevant actor to avoid harmful interference. The area wherein this notification and coordination will be implemented to avoid harmful interference is referred to as a 'safety zone'. A safety zone should be the area in which nominal operations of a relevant activity or an anomalous event could reasonably cause harmful interference. The Signatories intend to observe the following principles related to safety zones:
>
> (a) The size and scope of the safety zone, as well as the notice and coordination, should reflect the nature of the operations being conducted and the environment that such operations are conducted in;
> (b) The size and scope of the safety zone should be determined in a reasonable manner leveraging commonly accepted scientific and engineering principles;
> (c) The nature and existence of safety zones is expected to change over time reflecting the status of the relevant operation. If the nature of an operation changes, the operating Signatory should alter the size and scope of the corresponding safety zone as appropriate. Safety zones will ultimately be temporary, ending when the relevant operation ceases; and
> (d) The Signatories should promptly notify each other as well as the Secretary-General of the United Nations of the establishment, alteration, or end of any safety zone, consistent with Article XI of the Outer Space Treaty

Thus, the parties to the Accords can establish safety zones in the area of operations of an activity in order to avoid harmful interference with other activities conducted by actors involved in lunar operations. Safety zones are expected to be temporary, only for the duration of the mission. Notification and coordination on the size and scope of the area between the parties is essential to the practical operation of the safety zone. Lastly, information about safety zones is to be made available to the public.[53] All these element contribute to the application of the principle of 'due regard' as an operator could not undertake lunar mission and establishing a safety zone without pre-coordination, notification and taking precautions for do not create harmful interference with others. And so, having 'due regard' to others' activities, in a cooperative approach, appears to be a step towards the creation of safety zones.[54]

---

[53] Artemis Accords, 11.9.

[54] Christopher Johnson, 'The Space Law Context of the Artemis Accords', Part. 2 (SpaceWatch, 28 May 2020), https://spacewatch.global/2020/05/spacewatchgl-feature-the-space-law-context-of-the-artemis-accords-part-2/, accessed 21 February 2024.

Finally, the Artemis Accords acknowledge that further efforts will be required to "develop international practices, criteria, and rules applicable to the definition and determination of safety zones and harmful interference".[55]

5 Conclusion

The principle of 'due regard' is a core element of international law and space legal framework, illustrated by Article IX of the Outer Space Treaty. It represents a legal obligation for States Parties to the OST to abide by. A breach of 'due regard' may result in the international responsibility of States.

The new lunar exploration projects highlight the need to implement and to 'operazionalise' the principle of 'due regard' in order to avoid interference between the different operations that will be carried out, particularly on the Moon. Space actors conducting lunar activities should consider applying the principle as it represents a condition for the exercise of the freedom of outer space, as it limits and protects a State's or an operator freedom of action in outer space. The obligation of due regard is key for ensuring coexistence between, equally legitimate activities in a given area on a celestial body.

The Artemis Accords represent a concrete application of 'due regard' for space exploration, considering the sharing of information and scientific result, the protection of cultural heritage and the creation of safety zone to avoid interference, as well as the importance of consultation and international cooperation. Putting into practice these element will also contribute to the sustainability of space activities.

Some challenges remain in particular in the identification of affected States by a space activity, the effectiveness of consultations and information sharing which will depend on the openness of States to enter into diplomatic efforts. Consequently, space community need to develop consultation mechanisms on an inclusive and transparent base as space activities become more diverse with commercial entities adding complexity to space governance, considering also States that have not signed the Accords.

Lastly, the application of the 'due regard' principle under the Artemis Accords has yet to be clearly defined by States and operators, in particular the elements to provide in terms of data sharing, as well as the characteristics that consultations and safety zones will take. The Accords will serve as a precedent in State practice for future space exploration missions.

[55] Artemis Accords, 11.6, see also 11.11.

The Intergovernmental Panel on Climate Change and the Role of Satellites in Data Collections and Observations

Adithya Variath and Palak Taori

Abstract Future predictions of climate change indicate that greenhouse gas concentrations in the atmosphere will increase and many of the changes are irreversible. Even if greenhouse gas emissions from human activity are ceased, many parts of climate change and its effects will last for generations. With increasing temperature, there is a greater chance of sudden or permanent changes. The complimentary tactics of adaptation and mitigation can be used to lessen and manage the hazards associated with climate change. Significant emissions reductions over the coming decades can lower the risks associated with climate change, improve the chances of successful adaptation, lower the long-term costs and difficulties associated with mitigation, and support climate-resilient pathways for sustainable development. Recently, several significant and well-publicized initiatives have brought attention to the necessity of developing indicators that gauge global efforts to mitigate climate change. As technology advances, more Earth observation (EO) data must be collected to provide information useful for mitigating and adapting to climate change. The advent of high-performance computing and the information revolution are making it possible to stimulate the Earth system's evolution realistically at unprecedented scales. When evaluating the long-term cost-effectiveness of any agreed-upon policy initiatives, EO data will be crucial in light of shifts in the major environmental indicators. The Intergovernmental Panel on Climate Change notes that governments and the commercial sector are required to make educated decisions that can be obtained through dependable space-based observations. The chapter explores how the use of satellite-based remote sensing can contribute to the enhanced use of space technology to address relevant needs identified under the United Nations Framework Convention on Climate Change (UNFCCC). The chapter also delves into how international space laws and national space policies can work towards combating climate change.

Keywords Climate change · Satellite · IPCC · Space Law · Environmental Law · Satellite remote sensing

A. Variath (✉) · P. Taori
Centre for Research in Air and Space Laws, Maharashtra National Law University, Mumbai, India
e-mail: adithya.v@mnlumumbai.edu.in

H. Jahankhani et al. (eds.), *Space Law Principles and Sustainable Measures*, Space Law and Policy, https://doi.org/10.1007/978-3-031-64045-2_2

1 Introduction

Since the 1600s, when Greek philosophers first became interested in astronomy and space exploration, space technology has advanced significantly.[1] Both the Soviet Union and the United States worked on and enhanced satellites for many years following the launch of Sputnik. In an unmanned space race, the Soviet Union defeated the United States, and in 1961 they also sent the first person into space. The first person in space and the first to orbit the Earth was Yuri Gagarin. The first Space Station was launched into space by the Soviet Union shortly after Gargarin's first flight. The United States launched Apollo 11 in 1969. The three men who were inside the spaceship were the first astronauts to step foot on the moon when they arrived there. A significant milestone in space exploration, the moon landing paved the way for safer and more effective human space travel.[2] While commercial elements of space technology are well known, space holds value beyond the aspects of space tourism, space weaponization, outer space mining and telemedicine.[3] Space also holds within its ambit the capacity to play a vital role in changing the way humans are fighting against climate change. Many energy-saving and environmentally beneficial inventions are the result of space technologies. Satellite-based technologies are lowering carbon dioxide emissions from automobiles, wind turbines are becoming more efficient due to remote sensing technology, and solar cells are generating more energy due to developments in space technologies.[4]

---

[1] Moore, Patrick, and Robin Rees. "The History of Astronomy." In *Patrick Moore's Data Book of Astronomy*, 526–35. Cambridge: Cambridge University Press, 2011; Kulikovsky, P. G. "Studies on the History of Astronomy." *The British Journal for the History of Science* 2, no. 1 (1964): 84–89; Whewell, William. "History of Greek Astronomy." In *History of the Inductive Sciences: From the Earliest to the Present Times*, 107–8. Cambridge Library Collection—Philosophy. Cambridge: Cambridge University Press, 2010; Pottasch, S. R. "The History of the Creation of Astronomy & Astrophysics." *European Astronomical Society Publications Series* 49 (2011): 23–31; Mendes de Leon, Pablo, Hanneke van Traa, and Jessica Schechinger. "Space Law." In *The Practice of Shared Responsibility in International Law*, edited by André Nollkaemper and Ilias Plakokefalos, 453–78. Shared Responsibility in International Law. Cambridge: Cambridge University Press, 2017.

[2] Butler, Andrew M. "Beyond Apollo: Space Fictions after the Moon Landing." In *Solar Flares: Science Fiction in the 1970s*, 38–50. Liverpool University Press, 2012. Runcorn, S. K. "Space Missions to the Moon and Planets." *Highlights of Astronomy* 4, no. 1 (1977): 173–173.

[3] Gotlieb, A. E. "Nuclear Weapons in Outer Space." *Canadian Yearbook of international Law/ Annuaire canadien de droit international* 3 (1965): 3–35; Vanhullebusch, Matthias. "China and the Non-Weaponization of Outer Space: Towards a Relational Normativity." In *The Cambridge Handbook of China and International Law*, edited by Ignacio de la Rasilla and Congyan Cai, 379–94. Cambridge Law Handbooks. Cambridge: Cambridge University Press, 2024; Goedhuis, D. "Legal Aspects of the Utilization of Outer Space." *Nederlands Tijdschrift voor Internationaal Recht* 17, no. 1 (1970): 25–50; Byers, Michael, and Aaron Boley. "Space Mining." Chapter. In *Who Owns Outer Space?: International Law, Astrophysics, and the Sustainable Development of Space*, 130–85. Cambridge Studies in International and Comparative Law. Cambridge: Cambridge University Press, 2023; De Filippi, Primavera, and Andrea Leiter. "Blockchain in Outer Space." *AJIL Unbound* 115 (2021): 413–18.

[4] McCord, T. B., and D. P. Cruikshank. "Spectrophotometric Remote Sensing Of Planets and Satellites." *Symposium—International Astronomical Union* 96 (1981): 57–87; Leidner, Allison K., and Graeme M. Buchanan, eds. *Satellite Remote Sensing for Conservation Action: Case*

The need to deal with climate change has been more evident in recent years, as the scientific community has called for the necessity for prompt action.[5] The melting of glaciers in the Andes, Himalayas, and Alps is a result of global warming, a process caused by anthropogenic developmental goals. It is closely linked to industrialization, which has fueled this century, especially in terms of the demand for energy derived from fossil fuels, as noted in 2003 and 2007 by the Intergovernmental Panel on Climate Change (IPCC). Global warming is a direct result of the radiative forcing in the atmosphere, which is influenced by the emissions of greenhouse gases such as carbon dioxide (CO2), nitrous oxide (N2O), methane (CH4), and aerosols. The atmosphere and oceans are getting hotter as a result of this global warming. Heat absorption is the primary cause of the oceans' rising levels in this instance.

Governments realised that this issue needed to be addressed globally, so they created the United Nations Framework Convention on Climate Change as a means of facilitating international political discourse on the topic and the IPCC to provide the scientific framework for characterising the scope and severity of the problem. The United Nations organisations also fulfil other relevant functions. For instance, the World Meteorological Organisation (WMO) has long backed the creation and running of national meteorological offices or departments. These entities provide the data needed to track the Essential Climate Variables, which include salinity, water vapour, air and water temperatures, and sea ice.[6]

Climate change was highlighted by UN Secretary-General Ban Ki-Moon in 2009 as the "defining challenge of our time." The Intergovernmental Panel on Climate Change's Fourth Assessment Report emphasises the overwhelming body of scientific

---

Studies from Aquatic and Terrestrial Ecosystems. Cambridge: Cambridge University Press, 2018; Sorooshian, Soroosh, Phu Nguyen, Scott Sellars, Dan Braithwaite, Amir AghaKouchak, and Kuolin Hsu. "Satellite-Based Remote Sensing Estimation of Precipitation for Early Warning Systems." In *Extreme Natural Hazards, Disaster Risks and Societal Implications*, edited by Alik Ismail-Zadeh, Jaime Urrutia Fucugauchi, Andrzej Kijko, Kuniyoshi Takeuchi, and Ilya Zaliapin, 99–112. Special Publications of the International Union of Geodesy and Geophysics. Cambridge: Cambridge University Press, 2014; DeSaussure, Hamilton. "Remote Sensing by Satellite: What Future for an International Regime?" *American Journal of International Law* 71, no. 4 (1977): 707–24.

[5] William Boyd, Climate Change, Fragmentation, and the Challenges of Global Environmental Law: Elements of a Post-Copenhagen Assemblage, 32 U. PA. J. INT'l L. 457 (2010); Foote, Stephanie, and Jeffrey Jerome Cohen. "Introduction: Climate Change/Changing Climates." In *The Cambridge Companion to Environmental Humanities*, edited by Jeffrey Cohen and Stephanie Foote, 1–10. Cambridge Companions to Literature. Cambridge: Cambridge University Press, 2021; Dessler, Andrew E. "Is the Climate Changing?" Chapter. In *Introduction to Modern Climate Change*, 16–38. Cambridge: Cambridge University Press, 2021; Mittler, Daniel. "The Changing Ethics of Climate Change." *Ethics & International Affairs* 28, no. 3 (2014): 351–58; Millard, Stephen. "Macroeconomics and Climate Change." *National Institute Economic Review* 264 (2023): 1–7.

[6] Dawson, Edward. "Cat Art and Climate Change: Collecting in the Data Anthropocene." In *Collecting in the Twenty-First Century: From Museums to the Web*, edited by Johannes Endres and Christoph Zeller, 165–78. Boydell & Brewer, 2022; Mai, Laura. "Measuring It, Managing It, Fixing It? Data and Rights in Transnational and Local Climate Change Governance." *Transnational Environmental Law*, 2024, 1–23; Lawing, A. Michelle. "The Geography of Phylogenetic Paleoecology: Integrating Data and Methods to Better Understand Biotic Response to Climate Change." *Paleobiology* 47, no. 2 (2021): 178–97.

data demonstrating that climate change would jeopardise long-term prosperity and economic growth, as well as the very existence of the most vulnerable communities. Serious effects, including as rising sea levels, altered growing seasons, and a rise in the frequency and severity of extreme weather events like storms, floods, and droughts, are predicted by the IPCC.

Responding to climate change involves a two-pronged approach, "Reducing emissions of and stabilizing the levels of heat-trapping greenhouse gases in the atmosphere, i.e. mitigation, and Adapting to the climate change already in the pipeline, i.e. adaptation." Space technologies play a very critical role in both these approaches. The evaluation of a country's susceptibility to climate change is another application for satellites. Space observations are the best way to supplement ground-based measurements with the most recent data on land-use types and changes resulting from urban migration, poverty, conflicts, and population increase.[7] For instance, determining the need for adaptation measures will depend on how vulnerable coastal cities are. Space-based data can also be useful in evaluating the risk of crops in low-lying flood plains along the shore.[8] When it comes to risk assessment, space-based instruments provide an excellent platform for evaluating how vulnerable groups are exposed to various hazards in addition to climate change.

Global agreements on space debris, policies regarding the use of particular orbits, and, more recently, global navigation systems and space legislation are reached in the framework of the United Nations through the Committee on the Peaceful Uses of Outer Space (COPUOS). Member states acknowledged the value of space science and its applications in 1999 at the United Nations Conference on the Exploration and Peaceful Uses of Outer Space, as known as UNISPACE III.[9] The UNISPACE III led to the establishment of the International Committee on Global Navigation Satellite

[7] Orheim, Olav, and Baerbel Lucchitta. "Investigating Climate Change by Digital Analysis of Blue Ice Extent on Satellite Images of Antarctica." *Annals of Glaciology* 14 (1990): 211–15; Goodison, B. E., and A. E. Walker. "Use of Snow Cover Derived from Satellite Passive Microwave Data as an Indicator of Climate Change." *Annals of Glaciology* 17 (1993): 137–42; Stewart, Rebecca L., Matthew Westoby, Francesca Pellicciotti, Ann Rowan, Darrel Swift, Benjamin Brock, and John Woodward. "Using Climate Reanalysis Data in Conjunction with Multi-Temporal Satellite Thermal Imagery to Derive Supraglacial Debris Thickness Changes from Energy-Balance Modelling." *Journal of Glaciology* 67, no. 262 (2021): 366–84; Foote, Stephanie, and Jeffrey Jerome Cohen. "Introduction: Climate Change/Changing Climates." In *The Cambridge Companion to Environmental Humanities*, edited by Jeffrey Cohen and Stephanie Foote, 1–10. Cambridge Companions to Literature. Cambridge: Cambridge University Press, 2021; Intergovernmental Panel on Climate Change. *Climate Change 2007—Mitigation of Climate Change: Working Group III contribution to the Fourth Assessment Report of the IPCC*. Cambridge: Cambridge University Press, 2007.

[8] Gillings, Mark. "Embracing Uncertainty and Challenging Dualism in the GIS-Based Study of a Palaeo-Flood Plain." *European Journal of Archaeology* 1, no. 1 (1998): 117–44; Paul Gormley, W. "The Protection of the Earth-Space Environment: The Utilization of Remote Sensing Satellites to Implement Human Rights." *Proceedings of the ASIL Annual Meeting* 77 (1983): 2–2.

[9] Report of the Third United Nations Conference on the Exploration and Peaceful Uses of Outer Space, A/CONF.184/6, 1999, See "Appendix 1: S196 Proposal to UNISPACE III." *Symposium—International Astronomical Union* 196 (2001): 397–98.

Systems (ICG), a forum of providers and user communities of global navigation satellite systems (GNSS)[10] and the United Nations Platform for Space-based Information for Disaster Management and Emergency Response (UN-SPIDER). UNISPACE III also contributed to soliciting enhanced support for the UN Programme on Space Applications and for the regional centres, reshaping it into its present form.[11]

The United Nations Office for Outer Space Affairs (UNOOSA) has been working for many years to encourage the use of space-based information, supporting COPUOS in its capacity as a Secretariat to assist the political discussion among Member States.[12] In the context of climate change, UNOOSA's mission is to support the collection and subsequent use of data obtained via satellites to aid in modelling and comprehension of the phenomenon, identify adaptation and mitigation strategies, and monitor the long-term effects of these actions. In this context, the chapter explores how satellites are employed to track temperature variations, sea level rise, changes in polar sea ice and glaciers, emissions of greenhouse gases from industrial activities and deforestation, and other variables. The chapter uses IPCC reports to delve into how space technologies can mitigate/adapt to climatic changes.

2 Emerging Space Technologies, Innovation and Changing Climates/Climate Change

Climate change, driven by increasing greenhouse gas emissions, presents a significant danger to our planet's ecosystems, economy, and societal welfare. Although attempts have been made to reduce emissions, scientific forecasts indicate that many of the effects caused by climate change are permanent. The continuous buildup of greenhouse gases in the atmosphere guarantees that the effects will last for many

[10] Bonnor, Norman. "A Brief History of Global Navigation Satellite Systems." *Journal of Navigation* 65, no. 1 (2012); Brocklebank, D., J. Spiller, and T. Tapsell. "Institutional Aspects of a Global Navigation Satellite System." *Journal of Navigation* 53, no. 2 (2000): 261–71; Blanchard, W. F., and D. W. Broughton. "Institutional Requirements for a Global Navigation Satellite System." *Journal of Navigation* 48, no. 2 (1995): 249–55.

[11] Stevens, David, and Joerg Szarzynski. "Space-Based Solutions for Disaster and Emergency Medicine." *Prehospital and Disaster Medicine* 24, no. S1 (2009): s79–s79. The applications included satellite navigation, communications, disaster management, and meteorological forecasting for climate modelling. They also contributed to human well-being and sustainable development. In COPUOS, the topic of climate change was discussed at a special symposium held during the 46th session of the Scientific and Technical Sub-Committee, and more recently at the 52nd session, which was held in Vienna in June 2009.

[12] See also Taubenfeld, Howard J. "Consideration at the United Nations of the Status of Outer Space." *American Journal of International Law* 53, no. 2 (1959): 400–405; Simsarian, James. "Outer Space Co-Operation in the United Nations." *American Journal of International Law* 57, no. 4 (1963): 854–67; "United Nations: Convention on Registration of Objects Launched Into Outer Space." *International Legal Materials* 14, no. 1 (1975): 43–48; "United Nations Resolution on Legal Principles Governing Activities in Outer Space." *International Legal Materials* 3, no. 1 (1964): 157–59.

generations, even if emissions were stopped immediately.[13] This situation emphasizes the immediate requirement for thorough solutions that combine adaptation and mitigation measures to reduce the negative impacts of climate change. Adaptation entails coping with the unavoidable effects of climate change, whereas mitigation aims to reduce emissions to restrict additional warming.[14] Both methods are crucial components of a comprehensive strategy to address climate change, providing complementary solutions to tackle the issues caused by a warming planet. The present global warming phenomena are affected by variations in solar radiation caused by solar storms and different stages in the Sun's cycle.[15]

Satellite-based remote sensing is a vital tool in addressing the irreversible impact of climate change. Satellite data provide a comprehensive perspective on Earth's systems, offering significant insights into the complex interactions of environmental components that influence climate change. Advanced technologies, such as high-performance computers, allow for the gathering of large quantities of Earth observation (EO) data, which aids in the thorough monitoring of crucial environmental indicators.[16] The data are crucial for evaluating the effectiveness of mitigation measures and informing adaptation strategies in response to irreversible climate impacts. Recent efforts have highlighted the necessity of creating dependable indicators to measure worldwide advancements in mitigating climate change. Satellite-derived Earth Observation data are crucial for tracking environmental patterns and evaluating the effectiveness of policy measures.[17] Satellite data offer crucial insights into the changing climatic environment by monitoring sea level rise and alterations in plant patterns. By utilizing these technologies, governments and stakeholders may make better decisions to successfully tackle the issues of climate change. Global cooperation is crucial for optimizing the advantages of satellite-based remote sensing for addressing climate change through mitigation and adaptation. International cooperation is crucial to facilitate the unrestricted sharing of data and advance the establishment of standardized approaches for Earth Observation analysis.[18] Comprehensive space regulations, on both global and local scales, are essential to promote Earth observation missions related to climate change and to streamline the exchange of data.[19]

---

[13] Intergovernmental Panel on Climate Change (IPCC). (2018). Global warming of 1.5 °C. https://www.ipcc.ch/sr15/.

[14] United Nations Framework Convention on Climate Change (UNFCCC). (n.d.). https://unfccc.int/.

[15] Paul B. Larsen, Climate Change Management in the Space Age, 45 WM. & MARY ENV'T L. & POL'Y REV. 103 (2020).

[16] National Aeronautics and Space Administration (NASA). (n.d.). Earth observing system. https://eospso.nasa.gov/.

[17] United Nations Framework Convention on Climate Change (UNFCCC). (n.d.). https://unfccc.int/.

[18] United Nations Office for Outer Space Affairs (UNOOSA). (n.d.). https://www.unoosa.org/.

[19] European Centre for Space Law (ECSL). (n.d.). https://www.esa.int/About_Us/Law_at_ESA/European_Centre_for_Space_Law_ECSL.

The global community can fully utilize satellite technology to address the irreversible effects of climate change by promoting a collaborative environment. There have been various studies and papers that have spoken on the changing nature of the forests,[20] land use,[21] changing water cycles[22] etc. climate change is an existential threat that demands urgent and concerted action. The irreversible nature of many of its impacts underscores the importance of proactive adaptation measures and ambitious mitigation effortsmeasures and aggressive mitigation actions. Satellite remote sensing is a potent instrument for observing and comprehending the Earth's evolving climate, offering crucial data for making educated decisions. By utilizing these technologies and promoting global collaboration, we can address the obstacles posed by climate change and establish a more robust and enduring future for future generations.

3 Deconstructing Climate Change Adaptation and Mitigation

Adaptation and mitigation measures are crucial elements of the worldwide effort to address climate change. With growing greenhouse gas concentrations and permanent changes to the climate system, the importance of utilizing both techniques is becoming more evident. Adaptation is responding to the effects of climate change, whereas mitigation includes decreasing greenhouse gas emissions to restrict additional warming. Both techniques are crucial for reducing the negative impacts of climate change and enhancing resilience in susceptible populations and ecosystems.[23]

Implementing effective adaptation strategies can reduce the consequences of climate-related disasters such as floods, droughts, and heat waves, as well as safeguarding critical infrastructure and natural resources. For example, efforts to optimize water management and cultivate drought-resistant crops can strengthen the resilience of agricultural systems to evolving climatic circumstances. Furthermore, investing in infrastructure like seawalls and flood barriers can assist coastal towns in adjusting to higher sea levels and more frequent storm surges. Mitigation activities are essential for averting the most catastrophic outcomes of climate change by decreasing the release of greenhouse gases including carbon dioxide, methane,

---

[20] Li, H., G. Liu, C. Han, Y. Yang, R. Chen. Quantifying the Trends and Variations in the Frost-Free Period and the Number of Frost Days across China under Climate Change Using ERA5-Land Reanalysis Dataset. Remote Sens. 2022, 14, 2400.

[21] Basheer, S., X. Wang, A. A. Farooque, R. A. Nawaz, K. Liu, T. Adekanmbi, S. Liu. Comparison of Land Use Land Cover Classifiers Using Different Satellite Imagery and Machine Learning Techniques. Remote Sens. 2022, 14, 4978.

[22] Gnitou, G. T., G. Tan, Y. Hongming, I. K. Nooni, K. T. C. Lim Kam Sian. Resolution-Sensitive Added Value Analysis of CORDEXCORE RegCM4-7 Past Seasonal Precipitation Simulations over Africa Using Satellite-Based Observational Products. Remote Sens. 2022, 14, 2102.

[23] Data, The Intergovernmental Panel on Climate Change, https://www.ipcc.ch/.

and nitrous oxide.[24] Transitioning to renewable energy sources, improving energy efficiency, and enacting regulations to encourage sustainable land use are crucial mitigation options that may greatly decrease emissions.[25] The extensive use of renewable energy technologies like solar and wind power might help eliminate carbon emissions in the electrical sector and greatly decrease overall emissions.[26] Afforestation and reforestation programs can capture carbon dioxide from the atmosphere, aiding in reducing the effects of deforestation and land degradation.[27]

Adaptation and mitigation are separate techniques, however, they are tightly linked and support each other.[28] Mitigation actions can lessen the severity of climate change consequences and make adaptation efforts more feasible by decreasing emissions and preventing more warming.[29] Adaptation methods can enhance resistance to climate change, therefore decreasing the necessity for expensive mitigation actions in the future.[30] Combining adaptation and mitigation techniques can create synergies and enhance the efficacy of climate change response. Adaptation and mitigation have a role that goes beyond environmental factors and also includes social, economic, and political aspects. Marginalized groups, especially in low-income nations, are significantly impacted by climate change and frequently lack the means and ability

---

[24] Reilly, John M., and Francisco de la Chesnaye. "Mitigation of Greenhouse Gases." In *Human-Induced Climate Change: An Interdisciplinary Assessment*, edited by Michael E. Schlesinger, Haroon S. Kheshgi, Joel Smith, Francisco C. de la Chesnaye, John M. Reilly, Tom Wilson, and Charles Kolstad, 167–70. Cambridge: Cambridge University Press, 2007.

[25] Edenhofer, Ottmar, Ramón Pichs-Madruga, Youba Sokona, Kristin Seyboth, Susanne Kadner, Timm Zwickel, Patrick Eickemeier, et al., eds. *Renewable Energy Sources and Climate Change Mitigation: Special Report of the Intergovernmental Panel on Climate Change.* Cambridge: Cambridge University Press, 2011.

[26] Data, The Intergovernmental Panel on Climate Change, https://www.ipcc.ch/.

[27] Data, The Intergovernmental Panel on Climate Change, https://www.ipcc.ch/.

[28] Bizikova, Livia, Sarah Burch, Stewart Cohen, and John Robinson. "Linking Sustainable Development with Climate Change Adaptation and Mitigation." Chapter. In *Climate Change, Ethics and Human Security*, edited by Karen O'Brien, Asunción Lera St. Clair, and Berit Kristoffersen, 157–79. Cambridge: Cambridge University Press, 2010; SMITH, P., and J. E. OLESEN. "Synergies between the Mitigation of, and Adaptation to, Climate Change in Agriculture." *The Journal of Agricultural Science* 148, no. 5 (2010): 543–52; Zia, Asim. "Synergies and Trade-Offs between Climate Change Adaptation and Mitigation across Multiple Scales of Governance." Chapter. In *Adaptiveness: Changing Earth System Governance*, edited by Riyanti Djalante and Bernd Siebenhüner, 102–26. Cambridge: Cambridge University Press, 2021; Wiley, Lindsay F. "Mitigation/Adaptation and Health: Health Policymaking in the Global Response to Climate Change and Implications for Other Upstream Determinants." *Journal of Law, Medicine & Ethics* 38, no. 3 (2010): 629–39; Climate Change 2007—Mitigation of Climate Change: Working Group III contribution to the Fourth Assessment Report of the IPCC; Intergovernmental Panel on Climate Change (IPCC). "Climate-Resilient Pathways: Adaptation, Mitigation, and Sustainable Development." Chapter. In *Climate Change 2014—Impacts, Adaptation and Vulnerability: Part A: Global and Sectoral Aspects: Working Group II Contribution to the IPCC Fifth Assessment Report*, 1101–31. Cambridge: Cambridge University Press, 2014; Moellendorf, Darrel. "Treaty Norms and Climate Change Mitigation." *Ethics & International Affairs* 23, no. 3 (2009): 247–66.

[29] Data, The Intergovernmental Panel on Climate Change, https://www.ipcc.ch/.

[30] Data, The Intergovernmental Panel on Climate Change, https://www.ipcc.ch/.

to adjust.[31] Efforts to improve adaptation should focus on equity and social justice to help and empower the most vulnerable groups in dealing with the repercussions of climate change.[32] Mitigation measures should be governed by concepts of justice and fairness, including past emissions and the varied responsibilities of nations in tackling climate change.[33]

3.1 Principles of Space Law and Environmental Law

In the context of international space law, the principles of space law and general principles of environmental law will become a fundamental element in building international norms and agreements within climate change governance. This technique involves establishing a legally enforceable treaty that outlines general obligations for those who sign it, but deferring specific goals and particulars until future accords or domestic laws. The framework convention serves as a comprehensive text that establishes the values, aims, and governance norms of the treaty system, creating a basic structure for collective action. An illustrative instance of this approach in climate management is the UNFCCC, which aims to stabilize greenhouse gas levels to avoid harmful disruptions to the climate system. The Kyoto Protocol and the Paris Agreement build upon the promises of the UNFCCC by setting precise emission reduction targets and creating channels for global climate action. The transition from the UNFCCC to the Paris Agreement signifies concrete actions to combat climate change rather than merely political talk.[34]

Furthermore, the management of climate change now includes a wide range of stakeholders outside conventional governmental players. Multi-stakeholder alliances are increasingly recognized in global environmental governance as a way to tackle complex issues that are beyond the capacity of any individual sector. These stakeholders include government entities, non-governmental organizations (NGOs), companies, investors, and civil society groups. Non-party stakeholders, such as NGOs and business sector groups, are not official signatories to climate accords but are essential contributors to mitigation and adaptation activities. They contribute by offering knowledge, and resources, affecting policy choices, and encouraging cross-border collaboration. Market mechanisms supported by the Kyoto Protocol have increased incentives for collaboration among stakeholders, creating opportunities for creative solutions and collaborations. Multi-stakeholder engagement in global climate governance is now a key aspect of addressing the urgent issue of

---

[31] United Nations Framework Convention on Climate Change, May 9, 1992, S. Treaty Doc No. 102–38, 1771 U.N.T.S. 107.

[32] United Nations Framework Convention on Climate Change, May 9, 1992, S. Treaty Doc No. 102–38, 1771 U.N.T.S. 107.

[33] United Nations Framework Convention on Climate Change, May 9, 1992, S. Treaty Doc No. 102–38, 1771 U.N.T.S. 107.

[34] Yuan Dai & Shengli Jiang. Tackling Climate Change by Space Technology: The Pathway of the International Law, 19 US-CHINA L. REV. 439 (2022).

climate change, highlighting the importance of collaborative action for successful climate mitigation and adaptation.[35]

3.2 Soft Law and Space Law

Soft law mechanisms are being integrated into the international legal framework for outer space operations in addition to regular, legally enforceable treaties.[36] Non-binding measures like UN General Assembly resolutions and numerous guidelines and codes of conduct have benefits in dealing with sensitive and complicated problems such as those related to space technology and climate change.[37] Soft law is advantageous because it may let different parties reach an agreement without the constraints of formal legal requirements.[38] This is especially important in delicate regions where reaching agreement on enforceable agreements can be a time-consuming and difficult task. Moreover, the fast-paced changes in environmental concerns frequently require quick remedies that traditional legal methods may find difficult to offer.

Soft law can be essential in setting ethical standards and encouraging responsible behavior in the realm of space technology and climate change. Limitations on the use of advanced technology, especially in military contexts, can be set by non-binding legal instruments. This is particularly relevant due to the dual-use capabilities of several contemporary satellites, which may be utilized for both civilian and military functions.[39] Commercial satellites' capacity to collect and analyze data for military intelligence underscores the need for well-defined international guidelines to prevent abuse and promote peaceful utilization of space technology for addressing climate change.

---

[35] Bolin, Bert. "Key Features of the Global Climate System to Be Considered in Analysis of the Climate Change Issue." *Environment and Development Economics* 3, no. 3 (1998): 347–409; Peel, Jacqueline, Lee Godden, and Rodney J. Keenan. "Climate Change Law in an Era of Multi-Level Governance." *Transnational Environmental Law* 1, no. 2 (2012): 245–80; Coen, David, Julia Kreienkamp, and Tom Pegram. *Global Climate Governance.* of *Elements in Public and Nonprofit Administration.* Cambridge: Cambridge University Press, 2020.

[36] Freeland, S., & Zhao, Y. Rules of the "Space Road": How soft law principles interact with customary international law for the regulation of space activities. *Journal of Space Law* 44, no. 2 (2020): 405–432.

[37] Crawford, Mark. "Hard Choices, Soft Law: Voluntary Standards in Global Trade, Environment and Social Governance." *Canadian Journal of Political Science* 39, no. 1 (2006): 223–25; Mendes de Leon, Pablo, Hanneke van Traa, and Jessica Schechinger. "Space Law." Chapter. In *The Practice of Shared Responsibility in International Law*, edited by André Nollkaemper and Ilias Plakokefalos, 453–78. Shared Responsibility in International Law. Cambridge: Cambridge University Press, 2017.

[38] Yuan Dai & Shengli Jiang, Tackling Climate Change by Space Technology: The Pathway of the International Law, 19 US-CHINA L. REV. 439 (2022).

[39] Military looking for new ways to acquire and use commercial satellite data, Space News, https://spacenews.com/military-looking-for-new-ways-to-acquire-and-use-commercial-satellite-data/.

Soft law can promote the establishment of optimal methods for using space-based technology in environmental monitoring, resource management, and disaster response. Soft law mechanisms can promote international collaboration and collective action in tackling climate concerns by setting shared norms and promoting information exchange. It is important to recognize that soft law documents do not have the legal power of binding treaties. Their efficacy depends significantly on voluntary cooperation and the political determination of different nations. Soft law may have a substantial impact on international conventions, expectations, and behaviour despite not being legally binding. Over time, well-established soft law ideas may transform into customary international law, garnering wider recognition and even shaping the progression of subsequent obligatory agreements. The development of International space law principles can also align with environmental law principles in this context.

Adaptation and mitigation are critical measures that complement each other in tackling the difficulties of climate change. Enhancing resilience in at-risk populations and decreasing greenhouse gas emissions via adaptation and mitigation efforts can alleviate the effects of climate change and support sustainable development. To achieve a more sustainable and resilient future for everybody, space technology needs to combine adaptation and mitigation strategies while considering the social, economic, and political aspects of climate change. On the other side, the rapid advancement of technology and machinery in space also fueled the problem of the sustainability of space objects.[40] To mitigate the risk of space debris, space agencies and companies are developing technologies to track and avoid debris, as well as methods to remove it from orbit. These efforts are crucial to ensuring the safety and sustainability of space activities in the future. Efforts like United Nations guidelines on space debris mitigation are efforts from the perspective of international obligations on the States while exploring space.[41]

---

[40] Martinez LF, "Legal Regime Sustainability in Outer Space: Theory and Practice" (2019) 2 Global Sustainability 26; Jessup PC and Taubenfeld HJ, "Outer Space, Antarctica, and the United Nations" (1959) 13 International Organization 363.

[41] King MT and Blank LR, "International Law and Security in Outer Space: Now and Tomorrow" (2019) 113 AJIL Unbound 125; Liang QY, "Facing Seriously the Issue of Protection of the Outer Space Environment" in John A Simpson (ed), *Preservation of Near-Earth Space for Future Generations* (Cambridge University Press 1994); Simsarian J, "Outer Space Co-Operation in the United Nations in 1963" (1964) 58 American Journal of International Law 717; "United Nations Resolution on Peaceful Uses of Outer Space" (1964) 3 International Legal Materials 160; Freeland S and Gruttner E, "Critical Issues in the Regulation of Armed Conflict in Outer Space" in Suzannah Linton, Tim McCormack and Sandesh Sivakumaran (eds), *Asia–Pacific Perspectives on International Humanitarian Law* (Cambridge University Press 2019); "Agreement on Cooperative Activities in Outer Space: Soviet and American Institutions" (1963) 2 International Legal Materials 902; Katzenbach NDEB, "Sharable and Strategic Resources: Outer Space, Polar Areas, and the Oceans" (1959) 53 Proceedings of the American Society of International Law at its annual meeting (1921–1969) 206; Lang W, "Environmental Treatymaking: Lessons Learned for Controlling Pollution of Outer Space" in John A Simpson (ed), *Preservation of Near-Earth Space for Future Generations* (Cambridge University Press 1994); "International Cooperation in the Peaceful Uses of Outer Space Report of the Legal Sub-Committee of the U.N. Committee on the Peaceful Uses of Outer Space on the Work of Its Sixth Session" (1967) 6 International Legal Materials 1086; "United Nations Resolution on Legal Principles Governing Activities in Outer Space" (1964) 3 International

4 The Intergovernmental Panel on Climate Change and Earth Observation Data

Although lowering greenhouse gas emissions is crucial, the persistent climate change requires strong adaptation efforts to mitigate its negative impacts. Satellite remote sensing is a potent instrument for observing and comprehending the Earth system, offering crucial information for both mitigation and adaptation efforts. Yet, the lack of point-based weather measurements severely limits our comprehension of the Earth's evolving climate. This hinders the progress of understanding the Earth's climate system and developing accurate climate models for predicting future climate change, leading to significant uncertainty in future climate forecasts.[42]

Advancements in space technology are enhancing its capacity to assess, track, alleviate, and communicate climate change effects. There are over 2,000 operational satellites in space[43] at the moment. Over 20,000 satellites are anticipated to be orbiting in space soon.

The satellites will orbit in low-Earth orbit (LEO)[44] and feature advanced technologies to effectively combat climate change. Earth observation from space is beneficial as it is not constrained by territorial boundaries, unlike activities on Earth and in the atmosphere. Remote sensing is crucial for monitoring environmental conditions on land, sea, and in the atmosphere. Article 1 of the 1967 Outer Space Treaty (hereafter 'OST') states that:

> The exploration and use of outer space, including the moon and other celestial bodies, shall be carried out for the benefit and in the interests of all countries, irrespective of their degree of economic or scientific development, and shall be the province of all mankind. Outer space, including the moon and other celestial bodies, shall be free for exploration and use by all States without discrimination of any kind, on a basis of equality and in accordance with international law, and there shall be free access to all areas of celestial bodies. There shall be freedom of scientific investigation in outer space, including the moon and other celestial bodies, and States shall facilitate and encourage international co-operation in such investigation.

Remote sensing from space, such as monitoring maritime temperature and atmospheric conditions is an example of Article 1's vision of scientific investigation carried out for the interest of all countries. Remote sensing satellites can efficiently oversee and control the impacts of climate change on a worldwide scale. NASA and ESA pollution measurement satellites reported a substantial reduction in nitrogen

Legal Materials 157; "Outer Space: United Nations" (1975) 24 International & Comparative Law Quarterly 371.

[42] Wang, X. Remote Sensing Applications to Climate Change. Remote Sens. 2023, 15, 747. https://doi.org/10.3390/rs15030747.

[43] M. Bruno, Rocket Lab, Wallops Island Spaceport Expect Big U.S. Launch Growth, AVIATIONWK. (Dec. 18, 2019), https://aviationweek.com/defense-space/rocket-lab-wallops-island-spaceport-expect-big-us-launch-growth [https://perma.cc/4ZKF-XZ52].

[44] Kestutis Paulauskas, Space: NATO's Latest Frontier, NATO REV. (Mar. 13, 2020), https:// www.nato.int/docu/review/articles/2020/03/13/space-natos-latest-frontier/index.html [https://perma.cc/GEX5-9WGJ].

dioxide ("NO2") emissions from Chinese and Italian industrial activity between January 2020 and March 2020 as a result of COVID-19 quarantines imposed by the government. Given the increasing severity of climate change, the need for accurate data to guide efforts to reduce its impact and adjust to it is more crucial than ever. Earth observation data, especially from satellite-based remote sensing, provide a powerful method for monitoring and comprehending the intricate dynamics of our planet's climate system. With increasing greenhouse gas concentrations and irreversible changes occurring, the use of Earth Observation data in influencing policy choices and developing adaptive solutions is crucial. Adaptation involves adjusting to climatic impacts, while mitigation focuses on cutting emissions. These strategies work together to tackle the complex difficulties of climate change. Earth Observation data play a crucial role in these efforts by supplying vital information to evaluate environmental indicators, monitor advancements, and pinpoint vulnerable regions.

Knowledge of the climate system is based on observational data and model simulations.[45] Satellite remote sensing (SRS) is an essential method for analyzing the Earth's surface, subsurface, and atmosphere by utilizing sensors aboard satellites, especially geodetic satellites. Since 1959,[46] when radiometers on the Vanguard-2 spacecraft detected solar irradiance and cloud reflection, SRS has become a crucial research method for climate change studies.[47] Satellite remote sensing is now a fundamental aspect of efforts to address climate change, providing a broad perspective of Earth's processes on a large scale. Technological improvements, such as high-performance computing and the increase in satellite constellations, have led to a significant increase in the amount and resolution of Earth Observation (EO) data. This allows for more precise and rapid evaluations of environmental patterns. The statistics help assess the effectiveness of mitigation efforts and guide specific adaptation strategies to protect at-risk populations and ecosystems.

5 Reading the IPCC Reports and Contextualizing Space Technologies for Climate Change

The United Nations organisation responsible for evaluating the science around climate change is the Intergovernmental Panel on Climate Change. The Sixth Assessment Report, which includes a Synthesis Report and three contributions from Working Groups. The Working Group I, Working Group II, and Working Group III completed their contributions in August 2021, February 2022, April 2022, and

[45] 1. Overpeck, J. T., G. A. Meehl, S. Bony, D. R. Easterling. Climate data Challenges in the 21st Century. *Science* 331, 700 (2011).

[46] Yates, H. W. Measurement of the Earth Radiation Balance as an Instrument Design Problem. Applied Optics 16, 297–299 (1977).

[47] Li, J., M. H. Wang, Y. S. Ho. Trends in Research on Global Climate Change: A Science Citation Index Expanded-based analysis. Global and Planetary Change 77, 13–20 (2011).

March 2023, respectively, and the Synthesis Report was completed in August 2023.[48] According to the IPCC's Sixth Assessment Report, satellite-based observation is highlighted for its significant role in tracking about two-thirds of the key climatic variables, suggesting that satellite remote sensing is crucial for halting climate change.[49]

The 2021 Report emphasises how important satellites are for monitoring changes and enhancing climate prediction models. Improved and extended observational data records which have been available since the 2013 IPCC report provide more assurance for assessments of climate attribution.[50] The Climatic Change Initiative of ESA, which assists scientific teams in developing long-term datasets covering up to four decades for important climatic variables known as Essential Climate Variables, is expressly acknowledged in the study. Understanding the world's frozen regions has been transformed by remote sensing, especially in the vicinity of the poles where conditions make surface studies challenging.

These data monitor mass balance and elevation variations on thousands of isolated glaciers throughout the globe and evaluate their impact on sea level rise. Multiple mission observations, such as data from the EU's Copernicus Sentinel-1 mission, ESA's ERS-1, ERS-2, Envisat, and CryoSat missions, have proven essential for tracking changes to the ice sheets, which, along with glacier mass loss, are currently the main causes of global sea level rise. Updated and reconciled satellite estimates of the contributions of the ice sheet to sea level rise were provided by the ESA-sponsored Ice Sheet Mass Balance Inter-Comparison Exercise, or IMBIE.[51] The sea level has risen by 20 cm since 1901, and since 2006, the rate of growth has accelerated to a peak of 3.7 mm per year. The mean sea level may rise by an additional 28–101 cm by 2100, and it may rise by up to 2 m if the ice sheets become more unstable, depending on future carbon emissions. Sea-level change can be measured precisely thanks to satellite altimetry technology. In addition to the influence of thermal expansion, global measurements throughout the previous three decades are consistent with the contribution of sea level rise from ice losses.

The Copernicus Sentinel-6 Michael Freilich satellite was sent into orbit in November 2020 with the purpose of closely monitoring sea-level rise.[52] The satellite, which is the most advanced mission devoted to measuring sea-level rise, is expected to use the most recent radar altimetry technology to provide a new picture of ocean topography and extend the long-term record of sea-surface height measurements that started in 1992. Recent efforts have emphasized the necessity of dependable indicators to measure worldwide advancements in reducing climate change.

[48] Sixth Assessment Report, The Intergovernmental Panel on Climate Change, https://www.ipcc.ch/assessment-report/ar6/.

[49] Sixth Assessment Report, The Intergovernmental Panel on Climate Change, https://www.ipcc.ch/assessment-report/ar6/.

[50] Satellite data provide valuable support for IPCC climate report, European Space Agency, https://www.esa.int/Applications/Observing_the_Earth/Space_for_our_climate/Satellite_data_provide_valuable_support_for_IPCC_climate_report.

[51] Ice sheet Mass Balance Inter-comparison Exercise, http://imbie.org/imbie-3/.

[52] Sentinel-6 Michael Freilich Satellite, NASA, https://www.jpl.nasa.gov/missions/sentinel-6.

Satellite-derived Earth observation data provide abundant information useful for monitoring important environmental indicators and evaluating the effectiveness of policy measures. By utilizing these technologies, governments and the corporate sector can make better decisions to successfully tackle the issues of climate change. Climate change is a worldwide challenge that goes beyond boundaries and necessitates cooperative actions on a global level. Satellite remote sensing is becoming a vital instrument for comprehending and dealing with the intricacies of climate change. Global cooperation in space exploration and policy is crucial for maximizing the capabilities of satellite technology in addressing climate change. The IPCC highlights the significance of international collaboration in addressing climate change. Nations collaborate through efforts like the UNFCC to create strategies and policies to decrease greenhouse gas emissions and address the effects of climate change.[53]

Effective climate action needs precise and thorough environmental trend data, which is where satellite-based remote sensing becomes indispensable. Satellite Earth observation systems, like those run by NASA and ESA, offer extensive data on environmental factors including temperature, sea level rise, and land cover change. The data are crucial inputs for climate models, allowing scientists and policymakers to track changes in the Earth's climate system and evaluate the impact of mitigation and adaptation strategies. International cooperation is crucial for optimizing the use of Earth Observation data by allowing the unrestricted sharing of information and promoting the creation of standardized approaches for data processing and interpretation.

According to the Preamble of Outer Space Treaty, it "Reaffirms the importance of international cooperation in the field of activities in the peaceful exploration and use of outer space". The foundation of international cooperation in space exploration and policy is the necessity for well-defined and consistent space policies on both the global and national scales. The United Nations Office for Outer Space Affairs is pivotal in fostering global collaboration in space endeavours and aiding in the advancement of space law and policy. UNOOSA aims to guarantee that space exploration and usage benefit all of humanity and promote sustainable development and environmental preservation through programs like COPOUS.[54] International space laws are crucial for sustaining satellite-based Earth observation missions focused on monitoring and addressing the effects of climate change. These regulations regulate matters like data exchange, intellectual property rights, and space debris abatement, all crucial for the efficient functioning of satellite systems.[55] International cooperation in space policy can help alleviate geopolitical tensions and foster peaceful collaboration in space exploration, setting the groundwork for joint efforts on climate change. An instance of effective global cooperation in space-based climate monitoring is the Copernicus initiative, spearheaded by the European Union and ESA.

[53] United Nations Framework Convention on Climate Change, https://unfccc.int/.

[54] Committee on the Peaceful Uses of Outer Space (COPUOS). (n.d.). Home page, https://www.unoosa.org/oosa/en/ourwork/copuos/index.html.

[55] Space Law, United Nations Office for Outer Space Affairs, https://www.unoosa.org/oosa/en/ourwork/spacelaw/index.html.

The Copernicus program utilizes a group of satellites with sophisticated sensors to observe different elements of the Earth's ecosystem such as the atmosphere, seas, and land surfaces.[56] Copernicus offers free access to a wide range of Earth Observation (EO) data and climate indicators through its Climate Change Service, aiding worldwide initiatives to address and adjust to climate change.[57] Copernicus showcases the effectiveness of international collaboration in utilizing satellite technology for climate action by utilizing the knowledge and resources of several partner organizations.

Public–private collaborations are becoming more crucial in improving space-based climate monitoring and mitigation activities, alongside international collaboration. SpaceX, Planet Labs, and Maxar Technologies are engaged in creating and managing satellite constellations for Earth observation, offering additional data sources to government-run systems. Public–private partnerships can increase the temporal and geographical resolution of EO data, helping to overcome gaps in coverage and enhance our capacity to monitor and address climate change efficiently.[58] In the future, international cooperation and space policy will play a crucial role in combating climate change.[59] With the increasing severity of climate change effects, there will be a rising need for prompt and precise Earth Observation data to guide adaptation and mitigation plans.[60] Collaborating across borders and sectors allows stakeholders to optimize the advantages of satellite technology for climate action, contributing to the development of a more sustainable and resilient future for everyone.

International cooperation in space exploration and policy is crucial for progressing climate change via satellite-based remote sensing. Stakeholders may efficiently utilize satellite technology to monitor, reduce, and adapt to the consequences of

---

[56] Copernicus, https://www.copernicus.eu/en.

[57] Copernicus, https://www.copernicus.eu/en.

[58] Sarnoff, Joshua D., and Margaret Chon. "Innovation Law and Policy Choices for Climate Change-Related Public–Private Partnerships." Chapter. In *The Cambridge Handbook of Public–Private Partnerships, Intellectual Property Governance, and Sustainable Development*, edited by Margaret Chon, Pedro Roffe, and Ahmed Abdel-Latif, 245–88. Cambridge Law Handbooks. Cambridge: Cambridge University Press, 2018.

[59] Intergovernmental Panel on Climate Change (IPCC). "Climate-Resilient Pathways: Adaptation, Mitigation, and Sustainable Development." Chapter. In *Climate Change 2014—Impacts, Adaptation and Vulnerability: Part A: Global and Sectoral Aspects: Working Group II Contribution to the IPCC Fifth Assessment Report*, 1101–31. Cambridge: Cambridge University Press, 2014.

[60] See Golovko, V. A., M. Leppäranta, S. Kalliosaari, Y. U. S. Sedunov, and A. M. Volkov. "Ice Charting Based on Multispectral Satellite Data in the Baltic Sea." *Annals of Glaciology* 19 (1994): 121–25; Martinec, J., and A. Rango. "Interpretation and Utilization of Areal Snow-Cover Data from Satellites." *Annals of Glaciology* 9 (1987): 166–69; Wardinski, Ingo, and Erwan Thébault. "Modelling Internal and External Geomagnetic Fields Using Satellite Data." Chapter, 84–97; Remy, F., P. Mazzega, S. Houry, C. Brossier, and J. F. Minster. "Mapping of the Topography of Continental Ice by Inversion of Satellite-Altimeter Data." *Journal of Glaciology* 35, no. 119 (1989): 98–107; Irwin, T. J, S. Martin, and E. A. Munoz. "Determination of the Interannual Variability of the Bering Strait Transport from Satellite Data." *Annals of Glaciology* 17 (1993): 207–10.

climate change by increasing data sharing, standardizing procedures, and establishing public–private collaborations. Well-defined and logical space regulations on both the global and local scales are crucial for aiding these endeavors and guaranteeing that space exploration advantages everyone. Collaboration in space will be crucial for creating a sustainable and resilient future in the face of climate change issues.

6 Conclusion

Workable mitigation and adaptation strategies must be developed right away given the current state of climate emergency. The ability to combat climate change is increased by the introduction of groundbreaking technologies such as Space 2.0 systems and other advancements like artificial intelligence, the Internet of Things (IoT), 5G, and robotics. Satellite-based remote sensing has great potential to improve worldwide initiatives aimed at reducing and adjusting to climate change. Satellite technology may promote evidence-based decision-making and successful climate policy by offering fast and reliable data on environmental changes. Realising this promise necessitates ongoing investment in Earth Observation infrastructure, fostering international collaboration, and establishing strong space regulations. Satellite-based remote sensing, via coordinated efforts, can play a crucial role in our joint efforts to address the climate problem and create a more resilient and sustainable future for future generations. Cooperation between space agencies can facilitate the exchange of technologies to address both space exploration and climate change mitigation. For example, under the cooperative mission, NISAR, NASA and ISRO are collaborating to co-develop and deploy a satellite that would track groundwater supplies and offer alerts before volcanic eruptions. Additionally, it will track the rate of ice sheet melting and how it affects sea level rise. The Earth satellite will be the first dual-frequency radar imaging satellite, to monitor changes in the distribution of vegetation across national borders. Integration of international space law principles with environmental law principles can leverage space-based technologies and collaboration between nation-states to enhance climate change mitigation efforts and promote sustainable development on a global scale.

Quantum Revolution in Space: Enhancing Space Technology

Viduranga Shenal Landers

Abstract Quantum computing is an emerging field that has implications in numerous disciplines. This chapter explores the potential applications and implications of quantum technologies in space exploration. The chapter introduces quantum computing, highlighting its fundamental principles and supremacy over classical computing in addressing the challenges of space exploration. Quantum computing is often combined with other quantum technologies in order to maximize its beneficial effects. Quantum communication holds promise for secure satellite data transmission, quantum sensors hold promise for precise navigation and Earth observation, and quantum simulation improves the capabilities of spacecraft design and testing. Quantum computing offers novel solutions for trajectory optimization, space-debris monitoring, data processing and many more. In addition to the benefits and applications, this chapter highlights the legal, ethical, and sustainable considerations associated with quantum computing in space and discusses the importance of adopting sustainable measures and ensuring compliance with international space law in order to leverage this revolutionary technology in a safe and responsible manner.

Keywords Space · Quantum Computing · Communication · Quantum Sensing · Law

1 Introduction

Space exploration has always fascinated humanity. For ages, the vast expanse beyond Earth has been humanity's final frontier.

> Space, the final frontier… to explore strange new worlds; to seek out new life and new civilizations; to boldly go where no man has gone before. —Star Trek (Roddenberry, 1966)

Space has had connections with almost every scientific discipline since ancient times. From physics explaining the motion of planets and stars to biology shedding

V. S. Landers (✉)
University of Colombo School of Computing, Colombo, Sri Lanka
e-mail: vidurangalanders@gmail.com

H. Jahankhani et al. (eds.), *Space Law Principles and Sustainable Measures*, Space Law and Policy, https://doi.org/10.1007/978-3-031-64045-2_3

light on the origins of life on Earth, and even agriculture demonstrating the cultivation of plants in space aboard the International Space Station (ISS), each scientific discipline finds application in space, and all have benefitted from space exploration in some manner (Shea, 2023). The curiosity to understand the universe that we live in, including our cosmic neighborhood, has driven many ambitious exploration missions, fueled by technological advancements. From the initial astronomical observations using simple telescopes to the interplanetary missions of the Space Age, space exploration has transformed our understanding of the universe and our place within it. Today, advanced spacecraft and ground-based observatories provide details about planets, stars, galaxies, and the forces that influence them with an unprecedented level of accuracy. Nevertheless, the demand for more precise details remains high, and many further developments are anticipated in the coming years.

Despite the interest, space exploration has never been easy. Each and every initiative to explore space further has to deal with a unique set of problems. The extreme environments in space poses challenges in navigation, communication, and resource management and remote sensing has its own limitations. While there are a multitude of technical challenges related to space exploration, for the scope of this chapter, we are particularly interested in challenges related to information and communication technologies of space exploration missions. Tasks such as optimizing trajectories for fuel efficient travel or processing massive amounts of observational data, often strain the limits of classical computing due to their high computational demands. To list some challenges, trajectory optimization becomes intractable with numerous celestial bodies due to the complexity of solving many-body interactions, real-time image processing suffers from computational limitations due to extremely high data rates, spacecraft navigation and control demands continuous ground station communications due to errors from on-board sensors, and communication systems demand unbreachable security.

In the previous decades, computational limitations were perceived as tasks that could be efficiently solved with future generations of computers. However, as technological advancements in transistors are reaching their fundamental size limits, previous observations like Moore's law will no longer apply to computer chip manufacturing (Loeffler, 2022). Therefore, there is a pressing need to either develop highly efficient algorithms for every computational task or build computational models that are not limited by the physical size constraints of transistors. Quantum computing presents itself as one such computational model. Quantum computing offers the potential to address some of these complexities in classical computational limitations. Quantum computers utilize the principles of quantum mechanics, such as superposition and entanglement to perform calculations that are intractable for traditional computers (Arute et al., 2019). This increased computational power presents a new frontier for space exploration (Institute for Business Value (IBM), 2023).

Quantum technologies include more than just quantum computing. Quantum communication offers secure information transmission across vast distances, which is critical for safeguarding sensitive data (Gisin & Thew, 2007). Quantum sensors can enhance navigation capabilities and facilitate detailed surveys of celestial bodies due to their unprecedented precision (Bongs et al., 2023). Moreover, quantum simulation allows researchers to model complex phenomena, streamlining spacecraft design and accelerating the testing process.

With transformative potential comes the need for responsible governance. International space law is the body of law, consisting of treaties and agreements, that provide a framework for the peaceful and sustainable use of outer space (Garcia-Yarnoz, 2014). Key principles such as non-appropriation of celestial bodies, state responsibility for space activities, and the avoidance of harmful interference become even more pertinent with the advent of cutting-edge quantum technologies. Ensuring ethical deployment, equitability, and adherence to the law is crucial as quantum technologies are integrated into space exploration initiatives.

This chapter delves into the intersections of quantum technologies and space exploration. It begins with an overview of quantum computing (Sect. 2), followed by an in-depth examination of how quantum technologies closely associated with quantum computing can be utilized in the space sector with regard to computational tasks (Sect. 3), communication (Sect. 4), and sensing (Sect. 5). Each application sector is explained with the help of current research and drawing connections from recent advancements to future possibilities. The focus will then shift to the governance of quantum technologies in space (Sect. 6), analyzing the framework provided by international space law and proposing strategies to ensure harmony between technological progress and legal and ethical necessities.

2 Quantum Computing Fundamentals

This section offers a brief introduction to the field of quantum computing without delving deeply into its various subtopics. As such, it is advisable that this text not be used as a definitive reference on quantum computing principles, but rather as a broad overview to gain a high-level understanding of some of these principles. Following texts provide comprehensive information about quantum computing and related quantum technologies introduced in this chapter (Mermin, 2007; Nielsen & Chuang, 2011; Young, 2024; Wolf, 2021).

Quantum computers are devices that leverage the principles of quantum mechanics to attain an advantage over classical computers. Quantum mechanics is the mathematical framework that is used to explain physical phenomena at atomic and subatomic levels. As for the famous saying by Richard Feynman "If you think you understand quantum mechanics, you don't understand quantum mechanics": quantum realm completely contrasts with the classical physics that governs our everyday experience. It is completely governed by many counterintuitive behaviors of matter at the atomic and subatomic levels. At this scale, quantum phenomena like superposition, entanglement, and tunneling occur, paving the way for ground-breaking applications that push the boundaries of conventional technological limitations.

A quantum computer is a universal computing device, which is capable of performing any kind of computable computation including simulating a quantum computer itself. Quantum computing relies on some of the quantum mechanical principles like superposition, entanglement, and interference to achieve computational advantages over classical computers for specific kinds of problems. Analogous to classical bits in classical computation, the fundamental unit of quantum information is the qubit. Unlike classical bits, which definitively exist as either a 0 or a 1, qubits exist in a combination of both $|0\rangle$ and $|1\rangle$ states simultaneously by leveraging the quantum principle of superposition. Superposition is the ability for a quantum particle to be in multiple states or places at the same time. A superposition state $|\psi\rangle$ represents a probability distribution of the possible states and can be written as:

$$|\psi\rangle = \alpha|0\rangle + \beta|1\rangle$$

Here α and β are complex numbers that represent the probability distribution of the state between state $|0\rangle$ and $|1\rangle$. For example, if a qubit is in equal superposition between $|0\rangle$ and $|1\rangle$, it will collapse into state $|0\rangle$ or $|1\rangle$ with 50% probability when it is measured. In other words, if the qubit is measured 100 times, you will get 1 in ~ 50 of the measurements and 0 in ~ 50 of the measurements. The mentioned superposition state could be achieved with a Hadamard gate.

$H|0\rangle = \frac{1}{\sqrt{2}}|0\rangle + \frac{1}{\sqrt{2}}|1\rangle = |+\rangle$ or $H|1\rangle = \frac{1}{\sqrt{2}}|0\rangle - \frac{1}{\sqrt{2}}|1\rangle = |-\rangle$

Here, $H|0\rangle$ denotes the application of Hadamard gate (H) on a qubit register with a single qubit that is in state $|0\rangle$. Quantum computing is done by applying gates to quantum registers, similar to manipulating classical bits using classical logic gates using transistors. However, a quantum gate (U) is a unitary operation, which means:

$$U^{\dagger}U = UU^{\dagger} = I$$

Here the † symbol represents the conjugate transpose of U, and U is a unitary matrix relative to some basis. I denotes the relevant identity matrix. Single qubit gates, or gates that operate on a single qubit, can be visualized as rotations of the quantum statevector inside the Bloch Sphere (Fig. 1).

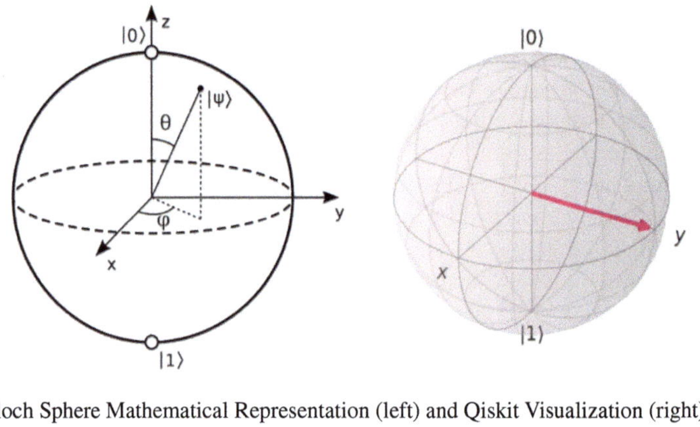

Fig. 1 Bloch Sphere Mathematical Representation (left) and Qiskit Visualization (right)

While qubits could exist in a superposition state, qubits need to be measured for their final outcome. Measurement collapses a superposition state into a definitive state. In other words, measurement strips off the added quantum value to the qubit and makes it similar to a classical bit. While quantum gates are reversible, measurements are irreversible operations. Based on our previous representation of the qubit state $|\psi\rangle = \alpha|0\rangle + \beta|1\rangle$, the measurement probability of state $|0\rangle$ is α^2, and measurement probability of state $|1\rangle$ is β^2. Here, α and β are under the constraint: $|\alpha^2| + |\beta^2| = 1$

Entanglement is one of the things that puzzled many early quantum physicists. Entanglement creates a connection between multiple qubits, causing their states to become inextricably intertwined. Which means that if one entangled qubit is measured, the states of its partners instantaneously collapse, regardless of the distance separating them. Two entangled qubit pairs are referred to as Bell pairs, which can be of four different forms denoted as $|\Phi^+\rangle$, $|\Phi^-\rangle$, $|\Psi^+\rangle$, $|\Psi^-\rangle$. Here, $|\Phi^+\rangle$ has the following form:

$$|\Phi^+\rangle = \frac{1}{\sqrt{2}}|00\rangle + \frac{1}{\sqrt{2}}|11\rangle$$

This means that if the measurement of one qubit results in state $|0\rangle$ the other qubit is automatically in state $|0\rangle$, and vice-versa.

At the current level of quantum computing, programming is done at qubit or gate levels. Similar to the early days of classical computing, where program instructions were given to computers by connecting wires and applying logic circuits, quantum computers are programmed at qubit level using quantum gates. A set of qubits called a qubit register is transformed from an initial quantum state to a preferred quantum state using quantum gates. Some of these gates include gates that are similar to those in classical, such as the NOT gate or X gate. However, there are many other rotation gates (Z, Y, RX, etc.) and gates such as Hadamard that allow us to make use of quantum features.

For example, the Hadamard gate transforms a qubit in state $|0\rangle$ to the previously discussed equal superposition between $|0\rangle$ and $|1\rangle$. Then applying a Controlled-X gate allows us to create an entangled Bell state, known as Bell state $|\Phi^+\rangle$. . These qubits are then measured to identify the outcomes (Fig. 2).

Quantum algorithms are carefully designed sequences of operations that harness the unique characteristics of superposition, entanglement and quantum interference. Here, I use the Bernstein-Vazirani problem to further explain how quantum computers are able to use their quantum features to attain an advantage over classical computers

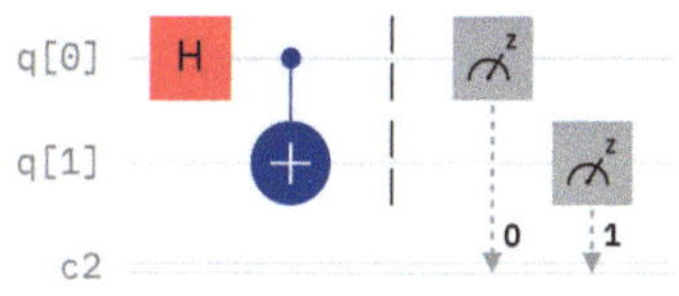

Fig. 2 Circuit to create a Bell state

for specific types of problems (Bernstein & Vazirani, 1997). The circuits were implemented on the IBM Quantum Platform using quantum assembly (QASM) language (IBM, 2024).

The algorithm tackles the problem of determining the function inside an oracle when the oracle behaves under certain constraints. In other words, Bernstein-Vazirani problem asks: given a function $f(x)$ that takes an input bitstring x of length n and produces a single bit output, and promises that there exists a bitstring s for which $f(x) = s \cdot x$ for all possible input bitstrings x of length n, is it possible to find the string s. In other words, given a function defined as $f : \{0, 1\}^n \rightarrow \{0, 1\}$, and the promise that there exists a string $s = s_{n-1} \ldots s_0$ such that $f(x) = s \cdot x$ for all $x \in \Sigma^n$, find the string s. Here, the binary dot product $s \cdot x$ is defined as:

$$s \cdot x = \begin{cases} 1 \textit{ if } s_{n-1}x_{n-1} + \cdots + s_0x_0 \textit{ is odd} \\ 0 \textit{ if } s_{n-1}x_{n-1} + \cdots + s_0x_0 \textit{ is even} \end{cases}$$

To solve the problem using a classical computer, one would need to evaluate the function for all possible input values. This requires $O(n)$ time, where n is the length of input bitstrings. In the quantum algorithm we pass all possible bit strings as a superposition state in a single evaluation and get the answer with a single query (Figs. 3 and 4).

Q-spheres are an extended version of the Bloch sphere which is used to showcase the quantum statevectors of multiple qubits. Here the circuit begins in state a, where all the qubits are in state $|0\rangle$. Then a series of Hadamard gates are applied to the oracle input quantum register to obtain all possible states into an equal superposition. Following that, the oracle is applied after which the states constructively and destructively interfere with each other resulting in the final state of $|0110\rangle$. Here, the first bit, which is the output register of the oracle, has no value to the measurement, and is used as an aid to construct the oracle circuit. Therefore, the states that represent as equal superpositions of $|00110\rangle$ and $|10110\rangle$ represent the final required bitstring $|0110\rangle$.

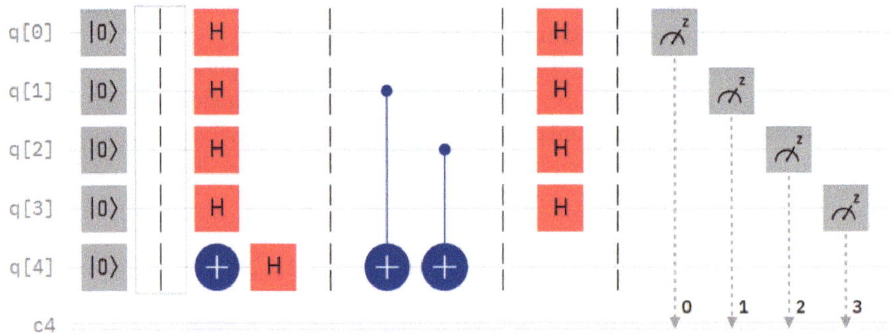

Fig. 3 Quantum Circuit for Bernstein Vazirani Oracle for "0110" string

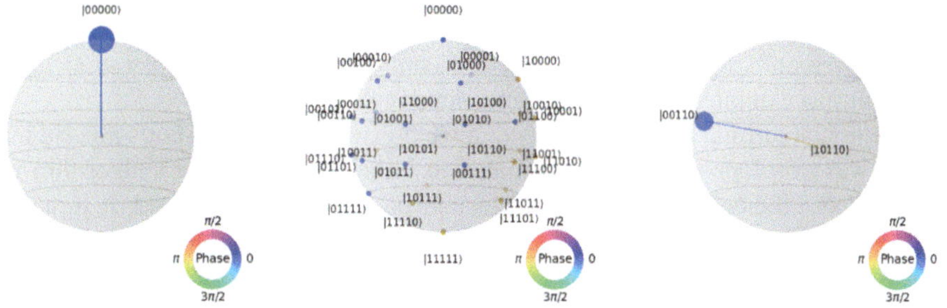

Fig. 4 Q-spheres of states: **a** left, **b** middle, **c** right

The purpose of the shown algorithm is to demonstrate the efficiency of quantum algorithms for given specific problems. While the Bernstein-Vazzirani algorithm has less practical use to real world problems, some of the landmark achievements in the field such Shor's algorithm and Grover's algorithm has many promising applications. Shor's algorithm has the theoretical potential to undermine commonly used encryption schemes by efficiently factoring large numbers (Shor, 1997). Grover's algorithm promises a quadratic speedup over traditional methods when searching unstructured databases, a capability that could revolutionize fields heavily dependent on rapid search capability (Grover, 1996).

Quantum technologies stretch far beyond just quantum computing. Quantum sensing harnesses the inherent sensitivity of quantum systems to measure incredibly minute changes in magnetic fields, gravity, and other physical quantities. Such precision holds promise for advancements in navigation, medical imaging, and the detection of valuable underground resources. Quantum communication employs entangled photons for the creation of unbreakable encryption methods, offering an unparalleled level of security in data transmission. Many countries are taking steps to accommodate a future 'quantum internet,' a globally distributed network of quantum devices seamlessly integrated for enhanced information processing and secure communication (European Commission, 2022).

Now that we understand what quantum computing is, so we'll move on to the potential applications of quantum computing and closely associated quantum technologies that are being explored for use in space exploration.

3 Quantum Advantage for Computational Tasks

The space missions we launch to expand the human understanding of the universe has to deal with complex problems, from navigating multi-body trajectories to surviving extreme environments. Solutions to many of these problems rely on the capabilities of computers to efficiently carry out complex tasks. In the previous section, we briefly discussed some quantum algorithms that hold promise for quantum supremacy over classical algorithms. In this section, we look at the potential of several quantum algorithms to efficiently solve some of the computationally demanding challenges, ongoing research for the use of quantum computing in space exploration, and potential space applications of quantum computing.

The first algorithm that we discussed in the previous section is Shor's algorithm, which was developed by mathematician Peter Shor in 1994. Shor's algorithm is designed to efficiently find the prime factors of a given large integer. It poses a significant threat to the security of widely-used public-key cryptography systems like RSA, which rely on the computational difficulty of factoring large numbers. Shor's algorithm leverages the principles of quantum superposition and interference to achieve an exponential speedup over the best-known classical factoring algorithms (Shor, 1997). While this algorithm may not have applications outside cryptography, it is an important milestone that proves the potential of quantum computing.

The next algorithm we discussed is Grover's algorithm, which was developed by Lov Grover in 1996. Grover's algorithm offers a quadratic speedup for unstructured search problems. In classical search, items should be checked one-by-one, requiring searches equivalent to half the search space on average. However, Grover's algorithm uses quantum superposition and amplitude amplification to find a marked item within a database of N items with roughly $\sqrt{N}$ queries (Grover, 1996). This has various applications in speeding up database searches, solving NP-complete problems, and in cryptography.

The two algorithms discussed above have been theoretically verified and practically demonstrated for simple versions due to the limitations of current quantum computers such as limited number of qubits, decoherence, and noise. For instance, Shor's algorithm was demonstrated on real quantum hardware back in 2001 (Vandersypen et al., 2001). There exist many other algorithms that are widely used for optimization problems, and have been demonstrated for various applications and using various quantum computer architectures.

Two of the core quantum algorithms that are used in many other algorithms are Quantum Phase Estimation (QPE) and Variational Quantum Eigensolver (VQE) algorithms. QPE is used to estimate the phase of an eigenvector of a unitary operator. It is a fundamental building block within Shor's algorithm for factoring, and is essential for solving linear systems of equations (Wong, 2022). VQE is a hybrid algorithm that helps to find the ground state energy of a chemical system described by a Hamiltonian. VQE uses a parameterized quantum circuit to prepare trial states whose energies are measured and then classically optimized (Tilly et al., 2022).

The Quantum Approximate Optimization Algorithm (QAOA) is a hybrid quantum–classical algorithm designed to address combinatorial optimization problems. QAOA aims to find approximate solutions to problems that can be expressed as finding the ground state of a Hamiltonian. It uses a parameterized quantum circuit that evolves according to the problem Hamiltonian, with the classical part optimizing the circuit's parameters. QAOA has a potential for applications in finance, logistics, and materials design (Farhi et al., 2014).

Quantum Machine Learning (QML) is an emerging field that explores the intersection of quantum computation and machine learning. QML aims to leverage the power of quantum computers to enhance machine learning algorithms in various

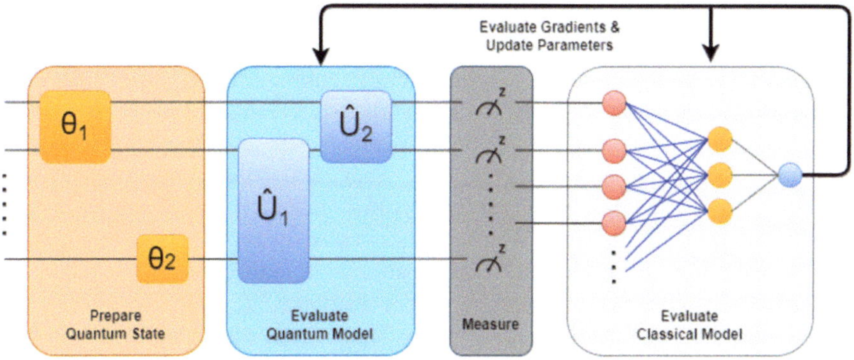

Fig. 5 Quantum Machine Learning Model

ways. These include potential speedups for tasks like linear algebra computations, the development of novel quantum-inspired machine learning models, and the exploration of quantum data that classical computers might find difficult to process (Acampora, 2019). Many of these algorithms include QPE, VQE, and QAOAs in their core and utilize similar strategies such as using a parameterized quantum circuit that is continuously evolving while a classical algorithm optimizes the parameters (Fig. 5).

One significant application of QML is for image classification. Quantum Image Classification algorithms explore the potential of representing images in quantum states and utilizing quantum operations for processing. Some techniques involve exploiting quantum superposition to encode multiple images simultaneously or using quantum walks for efficient image feature extraction. While still primarily theoretical, this field has the potential for future applications in image processing and computer vision.

Having discussed about several quantum algorithms and their potential to efficiently solve computationally demanding problems, it is time to address our most concerning question. Do these algorithms possess any practical relevance for the challenges within the space sector? If so, have there been any instances of their utilization, and what prospective applications may emerge in the future? The simple answer to these questions is: yes, quantum algorithms are capable of solving some of the challenges in the space sector and there are practical examples of their usage. So, let's first dive into the examples.

3.1 Scheduling Problems

The National Aeronautics and Space Administration (NASA) has been a pioneer in various space technologies. One of their first quantum applications was using quantum computing to solve an antenna scheduling task. Scheduling is a well-known NP-hard optimization problem that reduces to the Job Shop Scheduling problem

(Cozzolino et al., 2021). When it comes to space exploration, efficient scheduling of communication or tasks is a major challenge. This is especially true for satellite communication and other deep space spacecraft communication. NASA has taken the steps to use quantum computers to solve this issue (Guillaume et al., 2022).

NASA's Deep Space Network (DSN) has three main antennas, and they get a large number of requests to use them each week. Scheduling these requests is a complex process and takes up to four months to do manually. According to NASA, "Solving the DSN scheduling problem consists in placing as many tracks as possible within a finite time span (one week) while utilizing a limited amount of resources (antennas and equipment) and satisfying other constraints." To improve the efficiency of this process, NASA has explored the use of quantum annealing systems. They have used quadratic unconstrained binary optimization (QUBO) algorithms on D-Wave's quantum annealing systems (McGeoch et al., 2019) to produce more efficient schedules with more request satisfactions, more resource utilization, and less idle time.

A similar application was suggested by Kim et al., where they suggest the use of quantum optimization algorithms to reduce having overlapped monitoring areas among satellites in a satellite constellation. They have suggested the use of QAOA to address this challenge, and the problem of overlapping monitoring areas of the satellite constellation was mapped to a max-weight independent set problem (Kim et al., 2021). In another recent paper, Makarov et al. demonstrated the potential of quantum computing for the satellite image acquisition scheduling problem (Makarov et al., 2023). They have used QUBO algorithms on D-Wave's quantum annealing systems as well as a hybrid quantum solver to optimize the scheduling of image acquisition in Earth observation satellites.

These examples show that quantum computers can handle complex scheduling problems in the space sector. Space missions involve complex logistical planning for resources like fuel, power, and life support supplies. Classical computing often faces limitations in finding the most efficient ways to manage and distribute these limited resources, particularly for long-duration journeys. Furthermore, calculating routes for multi-body problems (Earth-Moon-spacecraft...) or those with gravitational assist maneuvers are classic bottlenecks. However, as quantum computers get better, they'll probably start playing a bigger role in solving such optimization problems (Desimone, 2019).

3.2 Material Development

Spacecraft often requires advanced materials that can withstand the harsh environments, and space mission design goes through rigorous testing of these spacecraft to ensure their ability to withstand harsh environments. Physical simulations are expensive, time-consuming, and may not fully capture the complex interactions at play in space. Some of the key fields where quantum simulations find applications are chemistry and material science. Quantum simulations can predict how materials behave under extreme conditions (radiation, temperature variations, microgravity), enabling the creation of more durable and lightweight spacecraft components.

Quantum computers are able to provide us with unprecedented insights into material properties by allowing us to efficiently simulate the behavior of atoms and molecules at the quantum level. This can lead to the discovery of novel materials with tailored characteristics like enhanced strength, conductivity, or resistance to radiation. Quantum computing could be used to screen vast libraries of potential materials, accelerating the identification of optimal candidates for spacecraft construction.

Importantly, the potential of quantum computing for materials simulation has moved beyond theoretical concepts. In 2021, a team of researchers at the Oak Ridge National Laboratory demonstrated for the first-time accurate material simulations that were verified with practical techniques (Humble, 2021). Since then, researchers have developed frameworks to simulate quantum chemistry models and have proven its success. These simulations are often done using models like the Ising model or Quadratic Unconstrained Binary Optimization (QUBO), and employing techniques such as quantum annealing (Buckeridge et al., 2023).

Moreover, quantum simulations can consider the complex mechanisms behind material degradation in the space environment. Understanding how extreme temperatures, radiation, and microgravity affect materials is vital for designing spacecraft that can successfully endure long-duration missions. By simulating these conditions at the quantum level, we can gain a deeper understanding of failure mechanisms, leading to the development of more resilient materials or protective coatings. This not only enhances spacecraft longevity but also minimizes the need for expensive and time-consuming physical testing, optimizing the overall development process.

3.3 Data Processing and Machine Learning

Space exploration missions generates immense datasets, from telescopic observations of distant galaxies to the torrent of data captured by Earth-imaging satellites. Analyzing this data for scientific insights, mission-critical decisions, and resource identification demands robust computational capabilities. Current state-of-the-art solution to analyzing large amounts of data is to employ Machine Learning (ML) models for pattern-based classification or recognition. Quantum Machine Learning holds the potential to revolutionize how we do machine learning, and promises the creation of highly efficient models.

As discussed previously, QML leverages the principles of quantum mechanics to enhance machine learning algorithms. This capability aligns well with the challenges faced in space exploration. For instance, deep space probes transmit terabytes of raw sensor data. QML algorithms could accelerate the identification of promising exoplanets, the detection of faint astronomical signals, or the classification of celestial objects, potentially uncovering phenomena overlooked by classical techniques (Schuld & Petruccione, 2021).

Here on Earth, hyperspectral imaging from satellites enables precise analysis of spectral signatures to identify ground composition, vegetation health, and even pollution sources. QML could empower researchers to develop advanced image processing techniques for hyperspectral data. Such applications could assist in a range of tasks, including image deblurring, resolution enhancement, or uncovering subtle spectral patterns previously masked by noise. These advancements could lead to various achievements in resource mapping, environmental monitoring, and even the detection of archaeological sites hidden beneath dense vegetation. More importantly, these techniques could be used for other space missions, which could be highly advantageous for resource mapping in space mining or other commercial space activities.

A recent study highlights the potential of QML in processing hyperspectral Earth imaging data. Researchers used quantum computers with limited number of qubits to develop a novel hyperspectral image restoration technology. In a recent paper, Lin and Chen claims to have demonstrated the viability of quantum hyperspectral remote sensing by restoring NASA's damaged hyperspectral images utilizing their newly developed hyperspectral quantum deep network called HyperQUEEN (Lin & Chen, 2023). This underscores the ability of QML to push the boundaries of conventional analysis and unlock valuable insights from existing datasets.

Several other works suggest the use of Quantum Neural Networks for hyperspectral image classification. In a study by Sebastianelli et al., the use of a circuit based hybrid Quantum Convolutional Neural Network is for classification of Earth observation images based on land use and land cover (Sebastianelli et al., 2022). In a recent conference publication, Zhang et al. suggests the use of Quantum-inspired Spectral Spatial Network for hyperspectral image feature extraction (Zhang et al., 2023). In both studies, the quantum solution was proven to be superior than current state-of-the-art image classification methods, which further confirms the potential of quantum computing.

The applications of QML in space exploration could extend beyond data analysis. As spacecraft become increasingly autonomous, onboard AI systems will be essential for real-time decision-making. QML could enable spacecraft to learn and adapt to changing environments, optimize navigation routes, and even identify potential hazards in previously unexplored territories.

While QML holds incredible promise, it's important to acknowledge that the field is still in its early stages. Significant advancements in both quantum hardware and algorithm development are needed to realize the full potential of QML in space applications. Nevertheless, the research and early successes indicate a transformative future where we can expect quantum machine learning to play a crucial role in expanding the horizons of space exploration.

3.4 Challenges

While the theoretical advantages of quantum algorithms for certain computationally demanding tasks are clear, several significant challenges must be overcome before their full potential can be realized in space applications. Today's quantum computers are classified as Noisy Intermediate-Scale Quantum (NISQ) devices, characterized by limitations in qubit count, high error rates, and short coherence times (Preskill, 2018). This current NISQ era presents hurdles for the execution of complex quantum algorithms that demand a large number of high-quality qubits for achieving significant computational speedups.

Additionally, quantum computers often rely on sensitive components and precise control mechanisms. The harsh space environment, with its radiation, extreme temperatures, and extreme vibrations during a spacecraft launch, challenges the development of robust quantum hardware suitable for space missions. Advances in radiation hardening, miniaturization, and fault-tolerant design are needed to make quantum computers viable for direct deployment in satellites or deep space probes.

Furthermore, qubits are inherently fragile and susceptible to errors introduced by environmental noise and decoherence. Complex error correction protocols are necessary for scaling quantum computers and achieving reliable computation. While error correction techniques are being actively researched, breakthroughs in this area will significantly influence the feasibility of powerful quantum algorithms in space applications (Roffe, 2019).

Finally, a major challenge is scaling existing quantum algorithms beyond demonstrations on small, well-defined problems. Real-world space exploration scenarios involve complexities like optimizing routes for large satellite constellations, processing massive amounts of sensor data, or simulating complex space conditions. This will demand further algorithmic development and potentially new strategies optimized for NISQ hardware.

4 Quantum Communication

Space exploration and the increasing reliance on space-based assets demand secure and robust communication channels. In the previous section, we discussed about the advantageous effects of quantum computing. However, in this section, we discuss about a technology that emerged due to the negative effects of quantum computing. Many traditional encryption methods that are used in communication, while computationally complex, are vulnerable to future quantum computers. Quantum algorithms such as Shor's factorization algorithm (Shor, 1997), Grover's search algorithm (Grover, 1996), and Regev's newly proposed factoring algorithm (Regev, 2024) has the ability to speed up key-less decryption of encrypted data. Among the sought-out solutions, the immediate solution is transitioning to quantum safe encryption protocols. The US National Institute of Standards and Technology (NIST) is currently leading the standardization of such post quantum cryptographic protocols ((NIST, 2016).

However, another solution exists. Quantum communication, offers a groundbreaking alternative, leveraging the principles of quantum mechanics to achieve unprecedented levels of security and enable novel networking paradigms in space. One of the most significant applications of quantum communication is Quantum Key Distribution (QKD). QKD is used to establish secret keys during a communication to facilitate a one time pad communication model. One time pads also known as Vernam ciphers are theoretically completely secure, and QKD offers a method to securely transfer the secret keys of a one time pad algorithm.

Many of these QKD protocols involve the transmission of photons to generate shared secret keys between two parties. Crucially, the quantum "no-cloning theorem" dictates that any attempt to eavesdrop on the transmission will unavoidably alter the quantum state, immediately alerting both communicating parties. This intrinsic security makes QKD an invaluable tool for protecting sensitive data transmitted between space crafts and ground stations, including mission control directives, classified payloads, and high-resolution scientific observations. There are two main categories of QKD protocols: measurement based protocols and entanglement based protocols. BB84 protocol, introduced by Bennett and Brassard in 1984, was the first QKD protocol to exist and it uses a measurement based method (Bennett & Brassard, 1984), and the first entanglement based QKD protocol was the E91 introduced by Ekert in 1991 (Ekert, 1991).

Beyond secure point-to-point communication, quantum technologies hold the potential to create vast quantum networks spanning Earth and space. Such networks could incorporate satellites equipped with entanglement generation and QKD capabilities, acting as secure nodes facilitating wide-area distribution of encryption keys. Moreover, quantum teleportation, the ability to securely transmit the quantum state of a particle over potentially vast distances, could lead to distributed quantum computing. This would allow space-based assets to access the processing power of multiple quantum computers, unlocking increased computational capability while maintaining data security.

Significant efforts have been made in developing space-based quantum communication systems. A pioneer in this field, China's Micius satellite, launched in 2016, demonstrated groundbreaking achievements in satellite-to-ground QKD over distances exceeding 1200 km (Yin et al., 2017). Moreover, Micius facilitated the world's first intercontinental quantum-encrypted video call between China and Austria, showcasing the real-world impact of this technology (Liao et al., 2017). In a successive research by Liao et al., an evaluation of how the demonstration of QKD using Micius satellite could be extended to create an ultralong distance global quantum network is discussed (Liao et al., 2018).

4.1 Challenges

While quantum communication holds the potential to revolutionize how we secure data and create distributed networks spanning Earth and space, significant challenges must be addressed to transition this technology from lab demonstrations to operational systems. A major challenge is the preservation of entanglement over long distances. Photons, the carriers of quantum information, are highly susceptible to decoherence, particularly when traveling through the vastness of space. Unlike terrestrial fiber optic networks, free-space transmission is necessary for space-based QKD. This, however, necessitates advanced techniques to mitigate signal degradation.

The effects of Earth's atmosphere pose a further challenge. Atmospheric turbulence and absorption introduce loss and distortions that can compromise the fidelity of the quantum signal. Mitigation strategies include the use of adaptive optics to compensate for atmospheric distortions, deploying high-altitude ground stations, or operating satellites above the densest layers of the atmosphere.

To realize robust and practical space-based quantum networks, scalability is another major focus area. Implementing QKD capable of supporting high data rates, establishing multiple simultaneous links for a network, and ensuring efficient key generation are essential for moving beyond proof-of-concept demonstrations. Research in areas such as optimizing photon distribution networks, multiplexing quantum channels to increase capacity, and developing high-performance key distillation techniques are crucial to address this scalability challenge.

5 Quantum Sensing

Remote sensing is an essential part of space exploration that helps us better understand the cosmos. From navigating the vast distances of the solar system to mapping Earth's resources and understanding astronomical mysteries, precise measurements are a foundational aspect of successful missions. Quantum sensing promises to revolutionize this field by harnessing the principles of quantum mechanics to achieve unprecedented levels of sensitivity in measuring physical properties such as time, magnetic fields, gravity, and acceleration.

At the heart of quantum sensing lies the ability to exploit quantum phenomena like superposition and entanglement. These allow quantum sensors to surpass the limitations of their classical counterparts. Applications of quantum sensing in space have a wide range.

We already discussed how quantum computing promise hyperspectral analysis that goes beyond what is possible with classical systems. In addition to analyzing existing hyperspectral data, quantum technologies can also be used to hyperspectral imaging. Quantum gravimeters could map Earth's subsurface with far higher resolution (Ménoret et al., 2018). Quantum magnetometers could capture faint electromagnetic signals. These measurements could lead to improving the search for aquifers, resource deposits, and potentially even predicting volcanic activity, revealing the composition of exoplanet atmospheres and potentially even offering clues about habitable worlds. This will revolutionize our ability to remotely observe Earth as well as the cosmos, including monitoring minute changes in Earth resources, pinpointing sources of pollution, and even detecting signs of space resources.

In the previous section, we extensively discussed how quantum computers could be used for hyperspectral image analysis. Recent studies suggest, that quantum technologies are not limited to the analysis of such images, but are capable of enhancing imaging itself. Some of the explored attempts to enhance imaging using quantum technologies include, sub-shot imaging, ghost imaging, and single photon imaging. A recent study by Zhang, England and Sussman utilizes entangled photons to enhance hyperspectral imaging by employing a modified quantum ghost spectroscopy system (Zhang et al., 2022). Realization of such technologies could be highly beneficial for future Earth observation satellites.

Quantum gravimeters are based on cold atom interferometers, which operate on the principles of quantum interference, allowing them to map minute variations in Earth's gravitational field with exceptional precision. For many years there have been concepts on using quantum gravimeters for space applications (Carraz et al., 2014). The European Space Agency's (ESA) proposed Space–Time Explorer and Quantum Equivalence Principle Space Test (STE-QUEST) mission aims to put a cold atom interferometer in orbit. It would test the fundamental 'Einstein Equivalence Principle' with extraordinary accuracy, potentially redefining our understanding of gravity and refining the very timekeeping systems that are crucial for navigation in space (European Space Agency, 2013). This could be a first step towards realization of quantum sensing in space.

Quantum gravimeters have many applications in space exploration. For example, mapping the distribution of water ice beneath the Martian surface is crucial for understanding the planet's potential for past or present life and for planning future human missions. Using orbiting quantum gravimeters capable of detecting minuscule variations in Mars' gravitational field, could reveal the presence and distribution of subsurface ice deposits, providing valuable insights into Martian geology and water resources.

Atomic clocks are known to have with their unrivaled accuracy and stability. Atomic clocks have been in existence for decades, and have been used in several space missions (Samuelson, 2019). Atomic clocks when paired with quantum accelerometers, offer a breakthrough in autonomous spacecraft navigation. One particularly compelling example of quantum sensing highlights the potential of quantum navigation for deep space missions. Currently, deep space navigation relies heavily on Earth-based communications, which can suffer from signal delays and degradation over vast distances. Quantum gyroscopes and accelerometers coupled ultra-accurate atomic clocks could equip a spacecraft to navigate autonomously for extended periods without the need of traditional Geosynchronous Positioning Satellite (GPS) systems.

Research institutions associated with NIST are actively developing ultra-precise timing links between Geosynchronous satellites using optical communication methods (Boutin, 2023). The advancements of these quantum navigation related sensors could enable spacecraft to maintain precise knowledge of their position and trajectory even when communication with Earth is impossible. This has profound implications for future missions exploring distant parts of our solar system where traditional navigation methods become impractical.

In addition to exploring the space, the unique environment of space is a valuable laboratory for testing fundamental laws of physics. Quantum sensors provide the tools to push these tests into new regimes. For instance, by measuring the gravitational redshift with unmatched precision, quantum clocks in orbit could test the limits of Einstein's theory of general relativity. This has implications for our understanding of gravity, potentially leading to the unification of general relativity with quantum mechanics, a long-sought goal of theoretical physics.

5.1 Challenges

We discussed how quantum sensors are advantageous for remote sensing, imaging and navigation. However, quantum sensors often rely on delicate mechanisms and complex calibration procedures. Translating the advantages of quantum sensing from the laboratory into fully operational space systems presents a number of challenges.

Firstly, the complexity of these devices often makes them delicate, posing a challenge for enduring the rigors of launch and the harsh environment of space. The harsh environment of space, with its extreme temperatures, vibrations, and radiation, can compromise the delicate components of quantum sensors. Advancements in engineering and materials science are needed to develop fault-tolerant designs that can withstand the rigors of launch and prolonged space missions.

Achieving accurate measurements with quantum sensors demands extreme isolation from environmental disturbances. Vibrations, temperature fluctuations, and even stray electromagnetic fields can introduce errors that compromise the fidelity of sensor readings. Designing spacecraft with effective shielding or incorporating active compensation mechanisms, all while minimizing added weight, is a crucial engineering challenge.

Finally, the sheer precision of quantum sensor data poses challenges in data processing, interpretation, and calibration. New algorithms and protocols are needed to transform raw sensor readings into accurate and usable information for scientific investigations or mission-critical operations. Developing these calibration and data analysis techniques is essential to harness the full potential of quantum sensors in space.

6 Legal, Ethical, and Sustainability Considerations

The emergence of quantum technologies in space applications, from quantum sensing to quantum communication, creates a new era of scientific exploration, commercial opportunities, and national security implications. However, realizing the full potential of these breakthroughs requires careful navigation of a complex landscape of governance, ethical considerations, and sustainability concerns. In this section we will discuss some of the aspects that needs attention as we progress with quantum advancements and anticipate to attain the previously discussed benefits with the use of quantum technologies for space applications.

As quantum technologies redefine what is possible in space, existing legal frameworks face evolving interpretations. The foundational Outer Space Treaty (OST), with its principles of non-appropriation and the equitable sharing of benefits derived from space activities, may need reassessment in the context of quantum-enhanced capabilities (UNOOSA, 1967). For instance, how can a nation that employs QKD for secure communications balance national security objectives with the commitment to cooperation and data sharing for scientific purposes? Similarly, in light of the Artemis Accords' focus on space resource utilization and transparency, proactive discussions are needed to address potential challenges posed by quantum technologies (NASA, 2020). If hyperspectral sensors powered by quantum analysis uncover space resources with unprecedented accuracy, can such discoveries remain in alignment with the Artemis Accord's call for the compliance with OST's articles II, VI, and XI that ensures common heritage of mankind?

The security landscape is also transformed by quantum technologies. Though QKD promises unbreakable encryption for certain applications, hybrid systems integrating quantum and classical components require re-evaluating a space asset's overall cybersecurity posture. New threat models must account for potential vulnerabilities arising from these integrated systems. Secure handling of data as it transitions between quantum-secured channels and legacy systems is crucial for safeguarding sensitive information. Additionally, dual-use concerns and non-proliferation become increasingly significant, as the same quantum breakthroughs powering scientific missions also hold clear military advantages. Disarmament treaties and control mechanisms will be vital in ensuring the peaceful utilization of space.

Quantum sensing and its potential for extremely high-resolution observations raise profound privacy and data protection issues. Ethical frameworks and legal principles that address responsible use need to keep pace with technological advances. Hyperspectral sensing capabilities could make previous notions of privacy protection inadequate, demanding fresh guidelines for both government and commercial satellite operations. Quantum Key Distribution (QKD) also raises concerns about surveillance and individual privacy when deployed across large scales. Defining responsible intelligence gathering aligned with individual rights and national security interests is essential.

On a different perspective, the advancements of quantum technologies could lead to an increased number of space missions. Due to the efficiency gains and precision advantages, small nations as well as current key space players will be interested in performing much more space missions in the future. This leads to an even bigger space debris problem than what we're facing today. In addition, the intersection of space debris mitigation and quantum computing offers a complex case study for updated liability frameworks. Existing space law conventions might need adjustments to account for the role of quantum-powered risk mitigation efforts in the event of accidents or collisions. Incentives within treaty structures should be developed to encourage quantum-aided debris cleanup or de-orbiting strategies, and frameworks should be reevaluated to accommodate a surge of space missions due to the quantum advancements.

Addressing these challenges hinges on several guiding principles. Collaboration between technologists, legal scholars, ethicists, and policymakers is imperative. Proactive standard development in areas such as hybrid system security and privacy-preserving use of quantum sensor data requires international partnerships to foster innovation while ensuring responsible use. Equity and sustainability must underpin new space treaties and technology sharing agreements, preventing a divide between nations with varying access to quantum technologies.

To proactively navigate the complexities, we need dedicated committees and working groups within bodies like the UN's Committee on the Peaceful Uses of Outer Space (COPUOS) specifically focused on quantum technologies in space. Organizations such as the International Telecommunication Union (ITU), experienced in regulating complex technologies, are pivotal for establishing standards for QKD systems, spectrum allocation, and addressing interference concerns. Finally, research funding agencies should incentivize projects that integrate legal, ethical, and social considerations alongside the core development of quantum space applications.

7 Conclusion

In conclusion, the quantum revolution holds extraordinary potential for transforming space exploration. By harnessing the principles of superposition, entanglement, and quantum algorithms, we can address longstanding limitations in trajectory optimization, data processing, communication security, and precision sensing. These advances will drive breakthroughs in scientific discovery, mission efficiency, and the safeguarding of critical space assets.

Yet, this transformation hinges not solely on technological achievement, but also on a parallel evolution of the governance around its use. Quantum communication networks offer secure data transmission but create potential friction points surrounding monitoring, and transparency. Quantum computation opens new frontiers while also raising concerns about liability, potential cybersecurity risks, and equity in the face of potential quantum divides between nations. Quantum sensors bring great sensitivity with associated complexities about privacy and responsible access to enhanced monitoring capabilities.

This chapter sought to explore not just the potential of quantum technologies but to initiate a crucial dialogue at the intersection of technology and space law. Revisiting principles enshrined in the Outer Space Treaty and Artemis Accords, and building upon them with foresight becomes critical to harnessing the quantum age. Studies on early QKD satellite experiments, quantum algorithms powering hyperspectral advancements highlight that this dialogue cannot remain theoretical. Proactive, internationally-guided efforts to develop standards, protocols, and safeguards are indispensable.

Looking ahead, it is essential to consider both the transformative potential and the need for proactive regulation in a 5–10-year time horizon. The potential for a space-linked quantum internet, with secure global communication and coordinated optimization of vast satellite constellations, could fundamentally reshape space operations. Advancements in error correction could lead to satellite-based quantum processing units boosting onboard AI for autonomous deep-space exploration and rapid analysis of Earth observation data. A fascinating possibility is the use of quantum computing, combined with the unique environment of space, to prototype novel materials impossible to test on Earth. Quantum computing will allow us to answer long-standing scientific questions, solve challenging logistical issues, and unlock new discoveries for the benefit of all. To achieve this responsibly demands fostering cross-disciplinary partnerships between technologists, lawyers, ethicists, and policymakers. The challenge and opportunity we face is that the principles embedded in our laws must now rise to match the complexity and capabilities of quantum technologies in space.

References

Acampora, G. (2019). Quantum Machine Intelligence. *Quantum Machine Intelligence, 1*, 1–3.

Arute, F., Arya, K., Babbush, R., Bacon, D., Bardin, J. C., Barends, R., Biswas, R., Boixo, S., Brandao, F. G., Buell, D. A., & Burkett, B. (2019). Quantum Supremacy Using a Programmable Superconducting Processor. *Nature, 574*(7779), 505–510.

Bennett, C. H., & Brassard, G. (1984). *Quantum Cryptography: Public Key Distribution and Coin Tossing*.

Bernstein, E., & Vazirani, U. (1997). Quantum Complexity Theory. *SIAM Journal on Computing, 26*(5), 1411–1473.

Bongs, K., Bennett, S., & Lohmann, A. (2023). Quantum Sensors Will Start a Revolution—If We Deploy Them Right. *Nature, 617*, 672–675.

Boutin, C. (2023, June 21). NIST Lays Groundwork for Future Ultra-Precise Timing Links to Geosynchronous Satellites. *NIST*. [Online]. https://www.nist.gov/news-events/news/2023/06/nist-lays-groundwork-future-ultra-precise-timing-links-geosynchronous (Accessed 11 March 2024).

Buckeridge, B. C. J., Warburton, P., Kendon, V., & Woodley, S. (2023). Quantum Computing and Materials Science: A Practical Guide to Applying Quantum Annealing to the Configurational Analysis of Materials. *Journal of Applied Physics, 133*(22).

Carraz, O., Siemes, C., Massotti, L., Haagmans, R., Silvestrin, P. (2014). A Spaceborne Gravity Gradiometer Concept Based on Cold Atom Interferometers for Measuring Earth's Gravity Field. *Microgravity Science and Technology, 26*, 139–145.

Cozzolino, C., Christiansen, B., Vallejos, E., Sorce, M., & Visk, J. (2021, December 15). Job Shop Scheduling. Cornell University Computational Optimization Open Textbook. [Online]. https://optimization.cbe.cornell.edu/index.php?title=Job_shop_scheduling (Accessed 20 February 2024).

Desimone, R. (2019, March). Quantum Computing Algorithms for Optimised Planning & Scheduling (QCAPS). [Online]. https://www.dwavesys.com/media/ru5atj5p/11_qcaps-qubits-europe-march-2019.pdf (Accessed 20 February 2024).

Ekert, A. K. (1991). Quantum Cryptography Based on Bell's Theorem. *Physical Review Letters, 61*(6), 661–663.

European Commission. (2022, November 25). Quantum Internet Alliance—Phase1. EU Publications Office. [Online]. https://doi.org/10.3030/101102140 (Accessed 15 February 2024).

European Space Agency. (2013). STE-QUEST Assessment Study Report. European Space Agency.

Farhi, E., Goldstone, J., & Gutmann, S. (2014). A Quantum Approximate Optimization Algorithm. arXiv preprint, vol. arXiv:1411.4028.

Garcia-Yarnoz, D. (2014, December 2). Space Law. United Nations Office for Outer Space Affairs. [Online]. https://www.unoosa.org/oosa/en/ourwork/spacelaw/index.html (Accessed 14 March 2024).

Gisin, N., & Thew, R. (2007). Quantum communication. *Nature Photonics, 1*(3), 165–171.

Grover, L. K. (1996). A Fast Quantum Mechanical Algorithm for Database Search. In *STOC '96: Proceedings of the twenty-eighth annual ACM symposium on Theory of Computing.*

Guillaume, A., Goh, E. Y., Johnston, M. D., Wilson, B. D., Ramanan, A., Tibble, F., & Lackey, B. (2022). Deep Space Network Scheduling Using Quantum Annealing. *IEEE Transactions on Quantum Engineering, 3*, 1–13.

Humble, T. (2021, August 4). *Quantum Computing Enables Unprecedented Materials Science Simulations.* U.S. Department of Energy. [Online]. https://www.energy.gov/science/ascr/articles/quantum-computing-enables-unprecedented-materials-science-simulations (Accessed 20 February 2024).

IBM. (2024). IBM Quantum Platform. IBM. [Online]. https://quantum.ibm.com/ (Accessed 12 February 2024).

Institute for Business Value (IBM). (2023). *The Quantum Decade.* IBM Corporation.

Kim, J., Kwak, Y., Jung, S., & Kim, J.-H. (2021). *Quantum Scheduling for Millimeter-Wave Observation Satellite Constellation.* arXiv:2108.00626 [eess.SY].

Liao, S. K., Cai, W. Q., Liu, W. Y., Zhang, L., Li, Y., Ren, J. G., Yin, J., Shen, Q., Cao, Y., Li, Z. P., & Li, F. Z. (2017). Satellite-to-ground Quantum Key Distribution. *Nature, 549*, 43–47.

Liao, S.-K., Cai, W.-Q., Handsteiner, J., Liu, B., Yin, J., Zhang, L., Rauch, D., Fink, M., Ren, J. G., Liu, W. Y., & Li, Y. (2018). Satellite-Relayed Intercontinental Quantum Network. *Physical Review Letters, 120*(3).

Lin, C.-H., & Chen, Y.-Y. (2023). HyperQUEEN: Hyperspectral Quantum Deep Network for Image Restoration. *IEEE Transactions on Geoscience and Remote Sensing, 61*, 1–20.

Loeffler, J. (2022, February 13). We Can't Make Transistors Any Smaller, is this the End of Moore's Law? *Interesting Engineering.* https://interestingengineering.com/innovation/transistors-moores-law (Accessed 16 March 2024).

Makarov, A., Franceschetto, M. M. T. E. O. G., Villar-Rodriguez, E., & Oregi, I. (2023). Optimization of Image Acquisition for Earth Observation Satellites via Quantum Computing. arXiv [quant-ph], arXiv:2307.14419v1.

McGeoch, C. C., Harris, R., Reinhardt, S. P., & Bunyk, P. (2019). Practical Annealing-Beased Quantum Computing. D-Wave Systems Inc.

Ménoret, V., Vermeulen, P., Le Moigne, N., Bonvalot, S., Bouyer, P., Landragin, A., & Desruelle, B. (2018). Gravity Measurements Below 10-9 g with a Transportable Absolute Quantum Gravimeter. *Scientific Reports, 8.*

Mermin, D. (2007). *Quantum Computer Science.* Cambridge University Press.

NASA. (2020, October 13). The Artemis Accords. [Online]. https://www.nasa.gov/wp-content/uploads/2022/11/Artemis-Accords-signed-13Oct2020.pdf (Accessed 16 March 2024).

Nielsen, M. A., & Chuang, I. L. (2011). *Quantum Computation and Quantum Information* (10th ed.). Cambridge University Press.

NIST. (2016, December 19). Announcing Request for Nominations for Public-Key Post-Quantum Cryptographic Algorithms. *NIST*. [Online]. https://csrc.nist.gov/news/2016/public-key-post-quantum-cryptographic-algorithms (Accessed 25 February 2024).

Preskill, J. (2018). Quantum Computing in the NISQ Era and Beyond. *Quantum, 2*, 79.

Regev, O. (2024). *An Efficient Quantum Factoring Algorithm.* arXiv [quant-ph], no. arXiv:2308.06572v3.

Roddenberry, G. (1966). Director, Star Trek: The Original Series. [Film]. Desilu Productions.

Roffe, J. (2019). Quantum Error Correction: An Introductory Guide. *Contemporary Physics, 60*(3), 226–245.

Samuelson, A. (2019, June 19). What is an Atomic Clock? *NASA*. [Online]. https://www.nasa.gov/missions/tech-demonstration/deep-space-atomic-clock/what-is-an-atomic-clock/ (Accessed 10 March 2024).

Schuld, M., & Petruccione, F. (2021). *Machine Learning with Quantum Computers* (2nd ed.). Springer.

Sebastianelli, A., Zaidenberg, D. A., Spiller, D., Le Saux, B., & Ullo, S. L. (2022). On Circuit-based Hybrid Quantum Neural Networks for Remote Sensing Imagery Classification. *IEEE Journal of Selected Topics in Applied Earth Observations and Remote Sensing, 15*, 565–580.

Shea, G. (2023, September 22). Why Go to Space? *NASA*. [Online]. https://www.nasa.gov/humans-in-space/why-go-to-space/ (Accessed 14 March 2024).

Shor, P. (1997). Polynomial-Time Algorithms for Prime Factorization and Discrete Logarithms on a Quantum Computer. *SIAM Journal on Computing, 26*(5), 1484–1509.

Tilly, J., Chen, H., Cao, S., Picozzi, D., & Setia, K. (2022). The Variational Quantum Eigensolver: A Review of Methods and Best Practices. *Physics Reports, 986*, 1–128.

UNOOSA. (1967). Treaty on Principles Governing the Activities of States in the Exploration and Use of Outer Space, Including the Moon and Other Celestial Bodies. United Nations Office for Outer Space Affairs. [Online]. https://www.unoosa.org/oosa/en/ourwork/spacelaw/treaties/introouterspacetreaty.html (Accessed 16 March 2024).

Vandersypen, L. M. K., Steffen, M., Breyta, G., Yannoni, C. S., Sherwood, M. H., & Chuang, I. L. (2001). Experimental Realization of Shor's Quantum Factoring Algorithm Using Nuclear Magnetic Resonance. *Nature, 414*, 883–887.

Wolf, R. (2021). *Quantum Key Distribution* (1st ed.). Springer Cham.

Wong, H. (2022). *Quantum Phase Estimation.* Springer. https://doi.org/10.1007/978-3-030-98339-0_28

Yin, J., Cao, Y., Li, Y.-H., Liao, S.-K., Zhang, L., Ren, J.-G., Cai, W. Q., Liu, W. Y., Li, B., Dai, H., & Li, G. B. (2017). Satellite-based Entanglement Distribution Over 1200 Kilometers. *Science, 356*, 1140–1144.

Young, P. (2024, January 16). An Undergraduate Course on Quantum Computing. [Online]. https://young.physics.ucsc.edu/150/phys_150_all.pdf (Accessed 10 February 2024).

Zhang, J., Zhang, Y., & Zhou, Y. (2023). Quantum-Inspired Spectral-Spatial Pyramid Network for Hyperspectral Image Classification. In *IEEE/CVF Conference on Computer Vision and Pattern Recognition (CVPR).*

Zhang, Y., England, D., & Sussman, B. (2022). *Snapshot Hyperspectral Imaging with Quantum Correlated Photons.* arXiv, no. arXiv:2204.05984 [physics.optics].

Quantum Reinforcement Learning: Advancing AI Agents Through Quantum Computing

Ved Seetohul, Hamid Jahankhani, Stefan Kendzierskyj, and Isuru Sandakelum Will Arachchige

Abstract The field of Quantum Reinforcement Learning (QRL) has gained much attention in recent years within the domain of artificial intelligence (AI) and machine learning (ML). With the widespread acceptance of AI agents across diverse industries, it is imperative to augment their potential, effectiveness, and flexibility. Quantum Reinforcement Learning (QRL) presents an innovative prospect that can overcome the constraints imposed by conventional AI algorithms and present revolutionary resolutions to intricate real-world predicaments. The significance of this research lies in its ability to address a primary obstacle in artificial intelligence, which is the development of efficient decision-making and learning mechanisms for AI agents. Current classical reinforcement learning algorithms have made significant advances, but they frequently encounter computational hurdles; particularly in situations where instantaneous decisions are crucial. QRL plays a significant role in revolutionising the design of systems that possess autonomous learning capabilities akin to human cognition. It holds immense promise with its potential for exponential acceleration that could transform how AI agents learn, adjust, and make decisions. Consequently, this research plays a critical role in examining the integration of quantum principles with reinforcement learning to unlock novel domains within artificial intelligence. It promotes the progression of quantum computing through the provision of pragmatic applications in the field of artificial intelligence. This allows for the creation of AI agents optimised by quantum computation, which can effectively perform a diverse range of tasks. It transcends the theoretical realm and fosters tangible solutions that can reshape industries and economies. However, it carries a great deal of risk if not monitored, managed, and controlled.

Keywords Quantum computing · Quantum reinforcement learning · QRL · AI · ML

V. Seetohul · H. Jahankhani (✉) · S. Kendzierskyj · I. S. Will Arachchige
Northumbria University, London, UK
e-mail: Hamid.jahankhani@northumbria.ac.uk

H. Jahankhani et al. (eds.), *Space Law Principles and Sustainable Measures*, Space Law and Policy, https://doi.org/10.1007/978-3-031-64045-2_4

1 Introduction

Throughout human history, individuals have consistently endeavoured to engage in the practice of pattern matching within datasets. Considering Kepler's to Newton's analysis of astronomical data, a multitude of mathematical techniques were devised. These techniques eventually became automated with the advent of digital computers in the mid-twentieth century. Advancements in computational capabilities have facilitated the analysis of increasingly intricate problems and the development of robust algorithms. These technological advancements have enabled us to handle a range of optimisations such as identifying films and job opportunities and effectively managing investments. Furthermore, they have also contributed to the discovery of novel pharmaceuticals that not only enhance human living conditions but also extend life expectancy. In the current age of information, as we continue to generate and store an increasing volume of data in our digital realm, the efficacy of algorithms is correspondingly enhanced. According to Grigorescu et al. (2020), in the pursuit of developing a highly advanced algorithm with autonomous learning capabilities, we are currently encountering a significant obstacle. As per Burg and Ausubel (2021), these algorithms exhibit a strong appetite for both data and computational resources, with the latter facing a fundamental constraint, as Moore's Law approaches its limits. The current approach being utilised involves employing multi-core, multi-thread computing to achieve enhanced processing speed through the parallelisation of algorithmic tasks (Dafir et al., 2021). In the context of computing, the emergence of parallel computing as a novel paradigm has garnered significant attention. Quantum computers, on the other hand, represent the epitome of parallel computing, operating in a manner that is conceptually and fundamentally distinct from classical parallel computers. The emerging devices, despite being in the early stages of development, have already demonstrated their potential significance in contemporary computing. Consequently, it is crucial that we actively pursue their further advancement, as this novel technology holds the potential to propel us beyond the current state of local optimisation.

1.1 Quantum Methodologies Background

Figure 1 provides an overview of the various methodologies employed in the quantum machine learning domain, encompassing a broad range of techniques based on the nature of the generated data, distinguishing between classical (C) and quantum (Q) data, as well as the type of device employed for data processing, whether classical (C) or quantum (Q).

The term "CC case" pertains to the integration of concepts from quantum computing into classical machine learning, thereby enhancing its capabilities (Sajwan & Jayapandian, 2019). An illustrative instance of this phenomenon can be observed in the field of many-body physics, where Tensor Networks have been

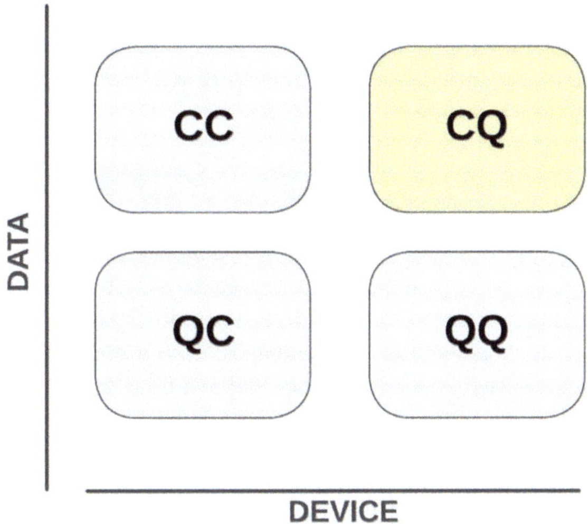

Fig. 1 Four approaches to quantum machine learning

employed. For instance, researchers have successfully utilised Tensor Networks to develop a novel framework for Markov decision processes, which serves as the fundamental basis for the field of Reinforcement Learning (Gillman et al., 2020). The QC case pertains to the inquiry into the potential of classical machine learning techniques to augment quantum computations. There exist numerous instances of this phenomenon, such as the acquisition of knowledge about quantum states (Montúfar, 2018). Additionally, a more recent development involves the application of reinforcement learning techniques to enhance the efficiency of quantum error correction codes.

There has been a growing interest in the community regarding the development of improved quantum error correction codes and fault-tolerant quantum computation techniques (Debenedictis et al., 2018). The QQ case stands out as a particularly intriguing example due to its comprehensive exploration of the quantum machine learning paradigm. One can interpret this phenomenon as the transmission of data produced by a quantum system into another quantum system. Alternatively, in a broader context, it involves employing a quantum computer to emulate the behaviour and evolution of a quantum system. As per Sanchez-Palencia (2018), the latter concept is known as quantum simulation. This approach has the potential to achieve exponential enhancements in computational speed. This approach has the potential to be applied in various contexts. As an illustration, let us consider the hypothetical scenario where nanorobots are employed within the human brain to potentially repair structures that have been damaged. The presence of authentic quantum structures known as microtubules within our brains suggests the potential existence of quantum computation. Consequently, it is plausible that a nanorobot could acquire knowledge by engaging with a sophisticated quantum environment (Tacchino et al., 2020). The concept of genuine quantum reinforcement learning emerges.

According to Telahun (2020), the CQ case pertains to the domain of quantum-enhanced machine learning. By now, numerous instances of such phenomena have gained significant recognition, encompassing a wide spectrum of applications. These include the utilisation of quantum enhancements in supervised learning, as well as the exploration of unsupervised learning challenges such as principal component analysis (Di Marcantonio, 2022). One potential application of this concept is the utilisation of a robot that engages with the physical environment conventionally but incorporates a quantum brain to enhance its cognitive capabilities. To ensure the functionality of this system, it is imperative to establish a conceptual framework for a quantum environment (Carneiro et al., 2021), which refers to the external realm in which a quantum agent engages. Traditionally, the field of reinforcement learning operates within environments that exhibit specific mathematical structures known as Markov Decision Processes. Consequently, to extend the classical environment to a quantum one, it is imperative to establish a mathematical framework. By possessing an understanding of a quantum environment, we are equipped to develop a quantum agent that operates according to the principles of quantum mechanics, as opposed to classical mechanics (Dunjko & Briegel, 2018). As a result, this quantum agent exhibits fundamentally distinct behaviour in comparison to its classical counterpart. The agent is now able to utilise the superposition principle to experiment with various actions in the environment and observe their outcomes. The environment provides feedback, which allows for the creation of parallel worlds, which are represented as a linear combination of multiple states, enabling simultaneous modifications to both the agent and the world states. The aspect that holds particular intrigue lies not in the superposition phenomenon itself. The utility of having multiple superposed states becomes limited when we are only able to observe a single classical state at a time through the act of measuring the superposition (Miguel-Ramiro et al., 2021). The concept being referred to is known as Holevo's Bound.

1.2 *Types of Quantum Learning*

Classical Reinforcement Learning: When comparing the approaches of supervised and unsupervised learning, which are commonly regarded as passive learning methods, it can be observed that reinforcement learning (RL) belongs to the category of interaction-based learning (Plaat, 2022). At a conceptual level, the learner engages with their surroundings, which they can perceive either completely or partially through an observation received after taking an action based on a predetermined policy (Haijiao et al., 2019). In the realm of Reinforcement Learning (RL), the learner is commonly referred to as an agent. This is because the agent can interact with its environment, whether it be in a simulated or real-world setting, based on its inherent capabilities (Abar et al., 2017). The primary objective of Reinforcement Learning (RL) is to acquire a policy by having an agent interact with an environment, such that the policy is optimised to maximise a reward that is specifically tailored to the problem at hand. However, the agent's objective is to discover the most effective strategy by exploring and evaluating various policies within the theoretical realm of all possible policies. The aim is to identify the policy that yields the highest anticipated total reward. Reinforcement Learning (RL) is grounded on the utilisation of Markov decision processes (MDPs) and the corresponding Bellman equation, which establishes its theoretical foundation as per Zhou, Gu, and Szepesvari (2021).

Quantum Supremacy: In 2019, Google successfully showcased the concept of quantum supremacy, providing a clear demonstration of its feasibility. This study demonstrates the viability of achieving full-scale quantum computing that surpasses classical supercomputers in terms of computational cost and time. Given the current limitations of classical computing power, the advancements being made in quantum computing are of great importance in the field of machine learning research. Quantum computers possess distinct capabilities that have the potential to revolutionise various areas of study, such as reinforcement learning, pattern recognition, and pre-trained natural language processing. The utilisation of quantum supremacy allows for the resolution of problems that were previously unsolvable by classical supercomputers. The fundamental source of strength in quantum computing lies in the distinction of the fundamental unit of information employed. In the field of quantum computing, qubits are employed as the fundamental units of information, replacing the conventional bits utilised in classical computing systems.

Parameter Utilisation: Quantum neural networks (QNNs) possess the capability to attain comparable performance as conventional neural networks (NNs) while utilising fewer parameters (Tacchino et al., 2020). This advantage can be attributed primarily to the phenomena of superposition and entanglement. The concept of superposition in quantum mechanics enables a quantum system to simultaneously exist in multiple states, thereby enhancing the representational capacity of quantum neural networks (QNNs) (Alchieri et al., 2021). Furthermore, entanglement serves as a reliable means of establishing connections between quantum states. Consequently, it enables the realisation of intricate and nonlinear feature representations while utilising a reduced number of parameters.

In the context of Quantum Neural Networks (QNNs), the parameter shift rule is employed as a technique for optimising parameters during the training process. Compared to the backpropagation algorithm commonly employed in traditional neural networks, the parameter shift rule presents more straightforward and simplified methodologies (Schuld & Petruccione, 2018). Hence, it is possible to expedite the training process in Quantum Neural Networks (QNNs).

Quantum Reinforcement Learning Algorithms: In the field of Quantum Machine Learning (QML), researchers have developed various approaches to address the challenge of stabilising the coherent function of the Quantum Processing Unit (QPU) through the utilisation of Machine Learning (ML) methods. Additionally, another set of approaches leverages the structure of a hybrid variational algorithm to achieve Machine Learning objectives within the QML framework. Reinforcement Learning (RL) is frequently employed to derive a solution for quantum control problems. These problems encompass tasks such as acquiring knowledge on quantum error correction strategies or developing control policies at a more granular level. Additional research has explored the concept of reinforcement learning (RL) in the context of a variational quantum algorithm (VQA). Although the topic of research mentioned is intriguing, our main focus here will be on the practical use of QRL algorithms for solving various tasks, whether they are in the classical or quantum domain. The existing body of research in the domain of Quantum Machine Learning (QML) has predominantly concentrated on the exploration of supervised and unsupervised learning methodologies.

Quantum-Inspired Approaches: The initial proposal for integrating reinforcement learning (RL) with a quantum routine is based on the concept of amplitude amplification, which is commonly employed in Grover-type search algorithms. Multiple quantum bit registers are utilised to incorporate the states and actions that are pertinent to the reinforcement learning system within an appropriate Hilbert space. The process begins with an initial state of equal probability distribution across all possible states. Subsequently, the amplitudes associated with states that yield higher rewards or align with the value function are selectively enhanced. The process of selecting an action is determined by Born's rule, which involves measuring the qubit register using the 'action basis.' The algorithm has been subject to independent investigation, separate from the study of Quantum Processing Units (QPUs), and has recently undergone additional advancements. Upon further examination, it is more appropriate to classify these initial iterations as a collection of Quantum-inspired Reinforcement Learning (QiRL) algorithms, which lack inherent capabilities for achieving quantum advantage. In recent times, there has been a transfer of the technique to the process of sampling from the experience replay buffer in the context of Q-learning.

2 Methods of Adopting Quantum Computing with Reinforcement Learning Frameworks

This section provides a thorough examination of the many strategies and tools that are methodically used to investigate the nexus between quantum computing and reinforcement learning (Gill et al., 2022). The research design of the study has been carefully crafted to address the particular requirements of the investigation, emphasizing its qualitative aspects. The procedure is the rigorous and careful collection of information from several sources that have been carefully chosen to provide comprehensive insights into the topic (Ahsan et al., 2022). Researchers examine both the theoretical as well as practical elements of the topics we study.

The investigators tried to examine both the theoretical foundations and the real-world applications. The ideal approach will rely on the extent to which the protocols align with the study's overarching objectives (Sun et al., 2018). The main objective is to offer a comprehensive analysis and rationale of the many techniques that have been used in the process of gathering data. These methods use a range of instruments and strategies, each serving a particular purpose in the hunt for relevant information (Lefebvre et al., 2019). When necessary, the study provides a detailed explanation of the moral principles that guide our interactions with people, as well as the openness that guides our participant selection procedure.

The legitimacy and dependability of the data are essential to scientific pursuits (Condie et al., 2022), providing clarity on the measures that researchers have taken to minimise bias and maintain consistency. An in-depth analysis of the ethical factors that influenced research endeavour is also provided, which focuses on important concepts, including informed consent, data confidentiality, and protecting one's privacy.

Rigorous quantum reinforcement learning research results examined the benefits of adopting quantum computing with reinforcement learning frameworks. The agents of quantum-enhanced artificial intelligence were subjected to extensive experimentation in which they were presented with a variety of scenarios designed to assess their adaptability, efficiency, and generalizability. The subsequent examination and elucidation of the data endeavour to provide a comprehensive understanding of the quantum advantage in situations involving reinforcement learning. The nuanced interplay between quantum entanglement, superposition, and reinforcement learning algorithms unfolds, revealing intricate insights into the quantum advantage. Additionally, the results are contextualized within existing literature, establishing connections to prior research while also delineating novel contributions (Martín-Guerrero & Lamata, 2022).

Leveraging the power of actual quantum processors from leading quantum computing providers constituted a crucial aspect of data collection. Quantum circuits and algorithms were executed on platforms such as IBM Quantum, Rigetti, and others, providing empirical data on the practical implications and efficiency gains of quantum-enhanced reinforcement learning. The utilization of quantum hardware allowed for a direct examination of how the quantum nature of computation impacts the learning dynamics of AI agents. Parallelly, classical computing systems played a pivotal role in benchmarking and comparative analysis (Duong et al., 2022). Classical reinforcement learning algorithms were implemented on traditional hardware to establish a baseline for performance metrics. This dual approach facilitated a nuanced understanding of the quantum advantage by discerning the improvements, if any, offered by quantum computing over classical counterparts. The data collection process was further enriched by drawing insights from existing scholarly works, research papers, and documented experiences in the realms of quantum computing and reinforcement learning. This comprehensive review of the literature helped contextualize the findings within the broader academic discourse, highlighting the novel contributions and advancements introduced by this dissertation. The multi-faceted approach to data collection, encompassing simulated environments, quantum processors, classical computing systems, and a thorough literature review, ensures a well-rounded and rigorous evaluation of the quantum reinforcement learning paradigm presented in this research study (Hu et al., 2019). The amalgamation of these diverse data sources strengthens the validity and applicability of the results, paving the way for informed discussions and insightful interpretations in the subsequent chapters.

2.1 *Thematic Analysis*

The following four themes of research analysis help to examine the integration of quantum principles with reinforcement learning and further unlock novel domains within artificial intelligence. It promotes the progression of quantum computing through the provision of pragmatic applications in the field of artificial intelligence and allows for the creation of AI agents optimised by quantum computation, which can effectively perform a diverse range of tasks.

2.1.1 Theme 1: Quantum Advantage in Learning Dynamics

The theme of "Quantum Advantage in Learning Dynamics" serves as the nucleus of the investigation into the intersection of quantum computing and reinforcement learning. At its core, this theme aims to unravel the unique ways in which quantum-enhanced models influence the learning trajectories and adaptive capabilities of AI agents (Glisic & Lorenzo, 2022). Quantum computing introduces a paradigm shift by leveraging principles such as superposition and entanglement. In the context of reinforcement learning, this theme explores whether these quantum phenomena confer a distinctive advantage in the speed and efficiency of learning dynamics. Preliminary analysis reveals instances where quantum-enhanced agents exhibit accelerated convergence in learning tasks compared to their classical counterparts (Elbadawi et al., 2021). The exploration extends to discerning patterns in the quantum states' evolution throughout the learning process, shedding light on how these states contribute to more rapid decision-making.

One key aspect under scrutiny is the quantum parallelism inherent in superposition, allowing quantum agents to simultaneously explore multiple possibilities. This concurrent exploration potentially accelerates the identification of optimal strategies, impacting the learning dynamics positively (Chen et al., 2020). Furthermore, the entanglement of quantum states may facilitate a more coherent and interconnected understanding of complex decision spaces, contributing to improved decision-making efficiency. Quantitative metrics, such as learning curves and convergence rates, serve as benchmarks for evaluating the quantum advantage (Mishra et al., 2021). By scrutinizing these metrics across various reinforcement learning tasks, from classical benchmarks to custom-designed environments, we aim to delineate the circumstances under which quantum-enhanced learning dynamics manifest most prominently.

This theme also invites a qualitative examination, seeking to uncover the underlying mechanisms through which quantum computing imparts its influence on learning. In doing so, we bridge the gap between theoretical expectations and empirical observations, offering insights into the practical implications of quantum advantage in the dynamic realm of reinforcement learning (Niu et al., 2019). In essence, the theme of Quantum Advantage in Learning Dynamics epitomizes our quest to unravel the distinctive contributions of quantum computing to the temporal aspects of AI agent development, laying the foundation for a comprehensive understanding of the transformative potential of quantum reinforcement learning. Continuing the exploration of the theme "Quantum Advantage in Learning Dynamics," we delve deeper into the intricacies of how quantum computing reshapes the temporal evolution of AI agents engaged in reinforcement learning tasks. The quantum advantage in learning dynamics extends beyond mere speed; it encapsulates the agent's ability to navigate complex decision spaces with unprecedented efficiency. Quantum mechanics introduces a layer of parallelism through superposition, enabling quantum-enhanced agents to simultaneously explore multiple paths in the decision-making process (Sarkar et al., 2021). This concurrent exploration potentially expedites the identification of optimal strategies, leading to accelerated learning dynamics. Moreover, the theme scrutinizes how this quantum parallelism adapts to diverse environments, assessing its robustness and efficacy across a spectrum of reinforcement learning scenarios.

Entanglement, another hallmark of quantum mechanics, adds a layer of sophistication to the learning dynamics. The interconnectedness of quantum states facilitates a more holistic comprehension of decision spaces, fostering coherence in the agent's learning process (Heimann et al., 2022). By entangling states relevant to successful strategies, quantum-enhanced agents may exhibit a more nuanced understanding of the environment, contributing to enhanced decision-making dynamics. Empirical data analysis is crucial in unpacking the nuances of quantum advantage. Learning curves, convergence rates, and temporal benchmarks offer quantitative insights into the speed at which quantum agents acquire knowledge and refine their decision-making policies (Agunbiade, 2022). Comparative analyses against classical models, particularly in scenarios where classical convergence is traditionally slow, provide a tangible demonstration of the quantum advantage in expediting learning dynamics.

Furthermore, the theme extends its inquiry into the adaptability of quantum-enhanced agents. Through dynamic environments and evolving tasks, the quantum advantage in learning dynamics should manifest across varying contexts, showcasing the versatility of quantum reinforcement learning in swiftly adapting to changing scenarios. Qualitatively, the investigation seeks to uncover the underlying mechanisms through which quantum advantage is expressed in learning dynamics (Niraula et al., 2021). This involves an examination of how quantum states evolve, converge, and diverge throughout the learning process. Insights gained from this qualitative exploration provide a nuanced understanding of the quantum advantage's tangible impact on the intricate dance of exploration and exploitation inherent in reinforcement learning.

2.1.2 Theme 2: Scalability and Resource Utilization

The theme of "Scalability and Resource Utilization" in the context of quantum reinforcement learning illuminates the pragmatic considerations associated with deploying quantum-enhanced AI agents on a broader scale. Scalability, a critical facet, assesses how well quantum reinforcement learning models adapt to increasingly complex tasks and larger datasets, signalling their potential for real-world applications. Quantum processors, while heralding a new era in computation, are not immune to limitations (Lamata, 2020). The theme navigates these challenges by investigating the scalability of quantum algorithms concerning the size and complexity of reinforcement learning problems. Does the quantum advantage persist as the dimensions of the problem space expand? Are quantum-enhanced agents capable of efficiently processing and learning from extensive datasets, or do they encounter diminishing returns?

Moreover, an integral aspect of this theme is the analysis of resource utilization. Quantum computing is known for its prowess in solving specific problems exponentially faster than classical counterparts, yet it comes at the cost of resource constraints, such as qubit coherence times and error rates (Kwak et al., 2021). Therefore, the theme scrutinizes how efficiently quantum processors handle the computational demands of reinforcement learning tasks, unraveling the delicate balance between computational speed and resource fidelity. Empirical data, stemming from experiments conducted on quantum processors, will be pivotal in assessing the scalability and resource utilization of quantum-enhanced agents. Comparative analyses against classical algorithms on tasks of varying complexities provide a basis for understanding the conditions under which quantum scalability yields tangible benefits (Martín-Guerrero & Lamata, 2021). This analysis also extends to the examination of resource trade-offs, delving into how quantum processors manage computational loads without compromising the fidelity of the learning process.

Continuing our exploration of the theme "Scalability and Resource Utilization" within the realm of quantum reinforcement learning, we delve deeper into the nuanced dynamics that unfold as we navigate the complex landscape of scaling quantum algorithms for AI applications. Scalability, in the quantum context, extends beyond the conventional notions of computational efficiency (Krenn et al., 2023). It encompasses the ability of quantum-enhanced agents to adapt seamlessly to larger problem spaces, demonstrating an inherent quantum advantage as dimensions expand. The theme meticulously scrutinizes whether quantum reinforcement learning models maintain their efficacy in tackling increasingly complex tasks and whether the observed quantum advantage persists or diminishes with growing computational demands. A pivotal aspect of this theme involves empirical assessments of quantum processors' performance under the strain of scalability (Saggio et al., 2021). Quantitative metrics, such as processing time and learning curve profiles, offer a lens into the practical implications of scalability, helping to identify the thresholds where the quantum advantage is most pronounced. Comparative analyses against classical algorithms provide benchmarks, revealing the conditions under which quantum scalability becomes a tangible asset.

Resource utilization emerges as a concomitant consideration, given the inherent constraints of current quantum processors. The efficient allocation of quantum resources, such as qubits and gates, becomes paramount in maintaining the fidelity of quantum computations. This aspect of the theme explores the delicate balance between computational speed and the integrity of the quantum state, shedding light on how resource limitations impact the overall learning process (Saggio et al., 2021). Beyond the technical intricacies, the theme prompts a broader reflection on the feasibility of integrating quantum-enhanced agents into real-world applications. Can quantum processors handle the data volumes and decision complexities inherent in practical scenarios? How do quantum algorithms scale when confronted with the diverse and dynamic challenges of real-world environments? These questions bring scalability and resource utilization to the forefront of discussions surrounding the practical implementation of quantum reinforcement learning. Qualitatively, the theme encourages exploration of potential strategies to enhance scalability and optimize resource utilization (Lockwood & Si, 2020). It seeks to uncover patterns and strategies that mitigate resource constraints, ensuring that quantum processors can harness their full potential in addressing the computational demands of large-scale reinforcement learning tasks.

2.1.3 Theme 3: Generalization Across Environments

The theme of "Generalization Across Environments" delves into the pivotal capability of quantum-enhanced AI agents to transcend the confines of specific training scenarios and generalize their acquired knowledge across diverse environments. In the realm of reinforcement learning, the ability to extrapolate learned strategies to new, unseen situations is a hallmark of intelligent decision-making, and this theme explores the extent to which quantum computing contributes to enhanced generalization (Wang et al., 2022). Quantum reinforcement learning models are subjected to a battery of diverse environments, ranging from classical benchmark scenarios to custom-designed tasks that simulate real-world complexities. The theme scrutinizes how well quantum-enhanced agents adapt to these variations, seeking patterns and trends that signify a superior capacity for generalization (Rajawat et al., 2023). The overarching question is whether the quantum advantage extends beyond specific contexts, enabling agents to navigate unfamiliar environments with efficacy.

Empirical analysis forms the backbone of this exploration, with a focus on evaluating the performance of quantum agents across a spectrum of environments. Learning curves, success rates, and adaptability metrics serve as quantitative indicators, offering insights into the robustness of quantum-enhanced agents in facing novel challenges (Khairy et al., 2020). Comparative analyses against classical reinforcement learning models provide a baseline for assessing the distinctive quantum contributions to generalization. The quantum principles of superposition and entanglement play a crucial role in this theme. Superposition enables quantum agents to explore multiple strategies simultaneously, potentially fostering a more versatile understanding of decision spaces (Schuld & Petruccione, 2021). Entanglement, on the other hand, contributes to a holistic comprehension of interconnected variables, enhancing the adaptability of quantum-enhanced agents to different environmental conditions.

Moreover, the theme prompts a qualitative exploration of the decision-making processes of quantum agents during generalization. By unraveling the quantum states' evolution across varied environments, we gain insights into the underlying mechanisms that empower quantum-enhanced agents to generalize effectively (Jhanwar & Nene, 2021). This qualitative lens complements the quantitative assessments, providing a comprehensive understanding of the interplay between quantum mechanics and the generalization prowess of reinforcement learning models. "Generalization Across Environments" stands as a critical theme in our quest to unravel the impact of quantum computing on reinforcement learning (Olivera-Atencio et al., 2023). It paints a vivid picture of how quantum-enhanced agents transcend the limitations of specific training contexts, showcasing their potential to become adaptable decision-makers in dynamically changing and diverse real-world scenarios.

Within the theme of "Generalization Across Environments," the exploration continues into the nuanced facets of how quantum reinforcement learning models demonstrate adaptability and cognitive flexibility across a spectrum of scenarios (Melnikov et al., 2023). This theme is foundational to understanding the practical implications of quantum computing, as the ability to generalize knowledge is a cornerstone of intelligent decision-making in dynamic and unpredictable environments. The empirical analysis under this theme involves subjecting quantum-enhanced AI agents to a series of environments that vary in complexity, dynamics, and contextual factors. Classical benchmarks serve as a reference point, allowing for a comparative evaluation of how quantum agents perform in novel, unseen settings. The theme places emphasis on quantifiable metrics such as transfer learning rates, transfer entropy, and transfer distances, offering a comprehensive evaluation of the generalization capabilities of quantum reinforcement learning models (Prajapati et al., 2023). A distinctive focus lies in uncovering the quantum advantage concerning generalization. Do quantum-enhanced agents showcase superior adaptability when transitioning from training to testing environments? Are there discernible patterns in the quantum states that correlate with successful generalization? These questions drive the exploration, guiding the analysis toward identifying instances where quantum computing confers a tangible benefit in navigating unfamiliar decision spaces.

2.1.4 Theme 4: Interplay of Quantum Mechanics and Reinforcement Learning Algorithms

The theme of the "Interplay of Quantum Mechanics and Reinforcement Learning Algorithms" delves into the intricate dynamics at the nexus of quantum computing principles and the algorithms governing reinforcement learning (Fikadu Tilaye & Pandey, 2023). This theme serves as the crucible where quantum mechanics, with its inherent complexities of superposition and entanglement, converges with the algorithmic foundations of reinforcement learning, offering a deeper understanding of the quantum advantage. At the heart of this theme is the exploration of how quantum mechanics, particularly superposition, amplifies the capabilities of reinforcement learning algorithms. Superposition, the ability of quantum bits (qubits) to exist in multiple states simultaneously, introduces a novel dimension to the decision-making process. Quantum-enhanced agents can explore multiple strategies concurrently, potentially expediting the identification of optimal policies and contributing to the agility of the learning process. This quantum parallelism challenges the sequential nature of classical algorithms, leading to an exploration of how this novel dimension impacts the efficiency and efficacy of reinforcement learning.

Entanglement, another quantum phenomenon, further complicates the landscape. The theme investigates how entanglement, the correlated state of multiple qubits, influences the decision-making synergy within reinforcement learning algorithms. The interconnectedness of entangled states may enhance the coherence of learned policies, providing a more holistic understanding of decision spaces (Cimini et al., 2023). By disentangling the impact of quantum entanglement on the dynamics of reinforcement learning, we unravel the potential for quantum-enhanced agents to navigate complex and interdependent decision environments more adeptly. Quantitative analysis within this theme involves scrutinizing the performance metrics of quantum-enhanced reinforcement learning algorithms in comparison to classical counterparts. Learning curves, convergence rates, and computational efficiency metrics offer insights into how the interplay of quantum mechanics influences the foundational aspects of reinforcement learning (Pramod et al., 2021). Comparative analyses across a spectrum of tasks provide a nuanced perspective on where and how quantum algorithms exhibit advantages, contributing to a refined understanding of the quantum-boosted learning dynamics.

Qualitatively, the exploration extends to unraveling the internal workings of quantum-enhanced agents during the learning process. By dissecting the evolution of quantum states as algorithms iterate, we gain insights into the subtleties of quantum advantage. This qualitative lens allows for the identification of patterns, correlations, and quantum phenomena that underpin the superior performance observed in certain reinforcement learning scenarios. This theme also prompts an investigation into the interpretability of quantum-enhanced reinforcement learning (Griol-Barres et al., 2021). Understanding how quantum states correlate with decision-making processes becomes crucial for practical implementation. Clarity on how quantum mechanics contributes to algorithmic outcomes is essential for building trust and facilitating the integration of quantum-enhanced AI agents into real-world applications.

3 Interpretation of Key Findings

The exploration of "Quantum Advantage in Learning Dynamics" has unearthed compelling insights into how quantum mechanics influences the temporal evolution of AI agents engaged in reinforcement learning. The key findings suggest a tangible quantum advantage, with quantum-enhanced agents showcasing accelerated learning dynamics compared to their classical counterparts. The observed quantum advantage in learning dynamics can be attributed to the quantum parallelism inherent in superposition. Quantum agents, leveraging superposition, engage in concurrent exploration of multiple strategies, enabling them to swiftly identify optimal policies. This parallelism expedites the learning process, allowing quantum-enhanced agents to navigate decision spaces with heightened agility that classical algorithms struggle to match (Yun et al., 2023). The interpretation here is that quantum parallelism contributes a dynamic and versatile dimension to the learning dynamics, presenting a promising avenue for the application of quantum computing in reinforcement learning scenarios.

However, the interpretation is not without nuance. The quantum advantage is context-dependent, with the magnitude of acceleration varying across different reinforcement learning tasks. This suggests that while quantum parallelism confers benefits, its efficacy is influenced by the nature and complexity of the learning problem. This nuanced understanding is critical for practical applications, guiding the deployment of quantum-enhanced agents in scenarios where the quantum advantage is most pronounced. Transitioning to "Scalability and Resource Utilization," the key findings shed light on the delicate balance between the scalability of quantum reinforcement learning models and the efficient utilization of quantum resources (Lockwood & Si, 2021). The findings indicate that quantum agents showcase scalability, adapting to larger problem spaces with relative ease. This scalability is a promising aspect, especially when considering the application of quantum computing to real-world scenarios with expansive datasets and intricate decision spaces.

However, the interpretation also unravels challenges related to resource utilization. Quantum processors, while scalable, face constraints such as qubit coherence times and error rates. The findings suggest that maintaining the delicate balance between computational speed and resource fidelity is crucial. Quantum processors need to efficiently allocate resources to handle the computational demands of reinforcement learning tasks without compromising the integrity of the quantum state. the interpretation underscores the dual nature of scalability and resource utilization in quantum reinforcement learning. Scalability opens doors to the application of quantum computing in broader and more complex scenarios, while resource challenges necessitate strategic optimizations for practical implementation. The findings illuminate a path forward, emphasizing the need for nuanced strategies that harness the scalability of quantum algorithms while mitigating the resource constraints inherent in current quantum processors (Wang et al., 2022). Together, these key findings provide a foundational understanding of how quantum mechanics shapes the learning dynamics and scalability of AI agents in reinforcement learning. The nuanced interpretations lay the groundwork for informed discussions on the potential and challenges of integrating quantum computing into the fabric of artificial intelligence.

Exploring "Generalization Across Environments" reveals key findings that underscore the adaptability and versatility of quantum-enhanced AI agents. The findings suggest that quantum reinforcement learning models exhibit a notable capacity to generalize knowledge across diverse environments, surpassing the limitations of specific training scenarios. This generalization prowess is a crucial attribute for real-world applications, indicating that quantum-enhanced agents can navigate and succeed in novel, unseen environments with efficacy. The interpretation of these findings emphasizes the transformative potential of quantum computing in the context of reinforcement learning (Rajawat et al., 2023). Quantum-enhanced agents, leveraging the principles of superposition and entanglement, showcase a superior ability to extrapolate learned strategies to unfamiliar contexts. This capability positions quantum reinforcement learning as a robust framework for applications that demand adaptability and resilience in dynamic and evolving scenarios. The implications extend to domains where traditional AI models might struggle to generalize effectively, such as in complex decision-making environments with evolving variables.

Moving to the "Interplay of Quantum Mechanics and Reinforcement Learning Algorithms," the key findings unveil a rich landscape where quantum mechanics profoundly influences the foundations of reinforcement learning. The findings highlight that quantum parallelism, rooted in superposition, accelerates the learning dynamics of quantum agents. This acceleration is a testament to quantum advantage, where quantum-enhanced algorithms explore multiple strategies concurrently, leading to faster convergence and adaptability (Schuld & Petruccione, 2021). Entanglement emerges as a crucial factor in the interplay, contributing to the coherence of learned policies. Quantum-enhanced agents, entangled at the quantum level, showcase a more holistic comprehension of decision spaces. This interconnectedness enhances the agents' ability to navigate complex and interdependent scenarios with a nuanced understanding, fostering superior decision-making. The interpretation of these findings underscores that the interplay between quantum mechanics and reinforcement learning algorithms is not merely additive but transformative (Jhanwar & Nene, 2021). Quantum principles introduce novel dimensions to the learning process, shaping the behaviour of agents in ways that classical algorithms cannot replicate. This interplay, evident in both the acceleration of learning dynamics and the enhanced adaptability of quantum agents, sets the stage for a paradigm shift in the landscape of intelligent decision-making.

3.1 Results of the Exploration

The different discoveries may be summed up by referring to how quantum mechanics was investigated and debated by the most recent technological advancements (Saggio et al., 2021). There was a discussion on the various facts and numbers that were particularly pertinent about how all of the adjustments needed to be made at various intervals of time. In addition, talks were held on the newly developed algorithms and how the new values needed to be incorporated into the existing ones. There were essentially four themes that had been produced throughout the whole of the scenario, and each of these four themes was tightly connected to the theme that had been generated. The quantum advantage that may be discovered in the dynamics of learning was one of these four topics that were taken into consideration here (Lockwood & Si, 2020). In addition, the scalability and the exploitation of resources were other factors that needed to be examined with a particular degree of attention. Researchers should take into account the generalizations that were made across the various settings.

4 Conclusion

Researchers who are investigating this topic should consider the geographical areas selected in a manner that is somewhat different currently. Several distinct places need to be focused on, and the audiences that are being targeted furthermore need to be unique. In this case, every one of the discoveries must be founded on facts and statistics, which can be described as being quite numerical (Sarkar et al., 2021). The distinct techniques would make it abundantly evident how the numerous results and outcomes would be comparable to one another, as well as how the findings that are offered by a secondary qualitative study may turn out to be quite different in comparison to what is often discovered. It is advised that the research gaps should be taken very seriously and that all of the prospective solutions that may solve these concerns should be worked on. Every one of the many study gaps, that would have been discovered in this specific investigation, has to be addressed and the prospective enhancements that may be produced to the topic need to be taken into consideration (Kwak et al., 2021). It is vital to investigate these themes thoroughly and concentrate on current research publications and necessary to have a grasp of the specific ways in which information may be obtained from a variety of sources to meet the numerous issues that are likely to arise about these situations. Also important is the use of well-known research journals, since they are the ones that are more likely to generate material that is authentic(Martín-Guerrero & Lamata, 2021). In addition, it is strongly suggested that ethical concerns should not be compromised in any way and all of the many academic ethics must be upheld since it is most probable that they are among how the information would be created on a more professional basis.

It is also advised that the researchers have an understanding of the fact that this is a topic field where the terminologies that are employed are rather difficult ones. This complexity has to be rectified and the numerous readers who are laypersons in the complete evaluation need to be respected in terms of how their learning requirements must be met more quickly (Niraula et al., 2021). Therefore, it is strongly suggested that all of the researchers carry out the kinds of investigations that would be rather simpler to comprehend and that the understanding of the data and figures of the study would not be among the more challenging ones.

There are many tangible benefits to continue from this chapter's findings and can build upon the findings and methodologies developed in this study to explore further applications and nuances of Quantum Reinforcement Learning. Practitioners can leverage these advancements to enhance their AI systems, making them more efficient, adaptable, and capable of handling complex decision-making scenarios. Industries can harness the power of quantum-enhanced AI for optimisation, risk management, and strategic decision-making, leading to competitive advantages and innovative breakthroughs.

References

Abar, S., Theodoropoulos, G. K., Lemarinier, P., & O'Hare, G. M. (2017). Agent-based modelling and simulation tools: A review of the state-of-art software. *Computer Science Review, 24*, 13–33.

Agunbiade, A. (2022, August 22). Quantum Computing & reinforcement learning: Partners in achieving artificial general intelligence. *SSRN Journal.*

Ahsan, M., Nygard, K. E., Gomes, R., Chowdhury, M. M., Rifat, N., & Connolly, J. F. (2022). Cybersecurity threats and their mitigation approaches using Machine Learning—A review. *Journal of Cybersecurity and Privacy, 2*(3), 527–555.

Alchieri, L., Badalotti, D., Bonardi, P., & Bianco, S. (2021). An introduction to quantum machine learning: From quantum logic to quantum deep learning. *Quantum Machine Intelligence, 3*, 1–30.

Burg, D., & Ausubel, J. H. (2021). Moore's Law revisited through Intel chip density. *PLoS ONE, 16*(8), e0256245.

Carneiro, J., Alves, P., Marreiros, G., & Novais, P. (2021). Group decision support systems for current times: Overcoming the challenges of dispersed group decision-making. *Neurocomputing, 423*, 735–746.

Chen, S. Y. C., Yang, C. H. H., Qi, J., Chen, P. Y., Ma, X., & Goan, H. S. (2020). Variational quantum circuits for deep reinforcement learning. *IEEE Access, 8*, 141007–141024.

Cimini, V., Valeri, M., Polino, E., Piacentini, S., Ceccarelli, F., Corrielli, G., Spagnolo, N., Osellame, R., & Sciarrino, F. (2023). Deep reinforcement learning for quantum multiparameter estimation. *Advanced Photonics, 5*(1), 016005.

Condie, C. M., Vince, J., & Alexander, K. A. (2022). Increasing polarisation in attitudes to aquaculture: Evidence from sequential government inquiries. *Marine Policy, 136*, 104867.

Dafir, Z., Lamari, Y., & Slaoui, S. C. (2021). A survey on parallel clustering algorithms for big data. *Artificial Intelligence Review, 54*, 2411–2443.

Debenedictis, E., Lentine, A. L., Marinella, M., Williams, R. S., Conte, T. M., & Gargini, P. (2018). Post Moore's law report (No. SAND-2018-5288). Sandia National Lab. (SNL-NM), Albuquerque, NM (United States).

Di Marcantonio, F. (2022). *The role of data in projected Quantum Kernels: The Higgs Boson discrimination.*

Dunjko, V., & Briegel, H. J. (2018). Machine learning & artificial intelligence in the quantum domain: A review of recent progress. *Reports on Progress in Physics, 81*(7), 074001.

Duong, T. Q., Ansere, J. A., Narottama, B., Sharma, V., Dobre, O. A., & Shin, H. (2022). Quantum-inspired machine learning for 6G: Fundamentals, security, resource allocations, challenges, and future research directions. *IEEE Open Journal of Vehicular Technology, 3*, 375–387.

Elbadawi, M., Gaisford, S., & Basit, A. W. (2021). Advanced machine-learning techniques in drug discovery. *Drug Discovery Today, 26*(3), 769–777.

Fikadu Tilaye, G., & Pandey, A. (2023). Investigating the effects of hyperparameters in quantum-enhanced deep reinforcement learning. *Quantum Engineering, 2023*(1), 2451990.

Gill, S. S., Kumar, A., Singh, H., Singh, M., Kaur, K., Usman, M., & Buyya, R. (2022). Quantum computing: A taxonomy, systematic review and future directions. *Software: Practice and Experience, 52*(1), 66–114.

Gillman, E., Rose, D. C., & Garrahan, J. P. (2020). *A tensor network approach to finite Markov decision processes.* arXiv preprint arXiv:2002.05185

Glisic, S. G., & Lorenzo, B. (2022). *Artificial Intelligence and quantum computing for advanced wireless networks.* Wiley.

Grigorescu, S., Trasnea, B., Cocias, T., & Macesanu, G. (2020). A survey of deep learning techniques for autonomous driving. *Journal of Field Robotics, 37*(3), 362–386.

Griol-Barres, I., Milla, S., Cebrián, A., Mansoori, Y., & Millet, J. (2021). Variational quantum circuits for machine learning. An application for the detection of weak signals. *Applied Sciences, 11*(14), 6427.

Haijiao, W. A. N. G., Zhen, Y. A. N. G., Wugen, Z. H. O. U., & Dalin, L. I. (2019). Online scheduling of image satellites based on neural networks and deep reinforcement learning. *Chinese Journal of Aeronautics, 32*(4), 1011–1019.

Heimann, D., Hohenfeld, H., Wiebe, F., & Kirchner, F. (2022). *Quantum deep reinforcement learning for robot navigation tasks*. arXiv preprint arXiv:2202.12180

Hu, F., Wang, B. N., Wang, N., & Wang, C. (2019). Quantum machine learning with D-wave quantum computer. *Quantum Engineering, 1*(2), e12.

Jhanwar, A., & Nene, M. J. (2021, August). Enhanced machine learning using quantum computing. In *2021 Second International Conference on Electronics and Sustainable Communication Systems (ICESC)* (pp. 1407–1413). IEEE.

Khairy, S., Shaydulin, R., Cincio, L., Alexeev, Y., & Balaprakash, P. (2020, April). Learning to optimize variational quantum circuits to solve combinatorial problems. In *Proceedings of the AAAI Conference on Artificial Intelligence* (Vol. 34, No. 03, pp. 2367–2375).

Krenn, M., Landgraf, J., Foesel, T., & Marquardt, F. (2023). Artificial intelligence and machine learning for quantum technologies. *Physical Review A, 107*(1), 010101.

Kwak, Y., Yun, W. J., Jung, S., Kim, J. K., & Kim, J. (2021, October). Introduction to quantum reinforcement learning: Theory and pennylane-based implementation. In *2021 International Conference on Information and Communication Technology Convergence (ICTC)* (pp. 416–420). IEEE.

Lamata, L. (2020). Quantum machine learning and quantum biomimetics: A perspective. *Machine Learning: Science and Technology, 1*(3), 033002.

Lefebvre, C., Glanville, J., Briscoe, S., Littlewood, A., Marshall, C., Metzendorf, M. I., Noel-Storr, A., Rader, T., Shokraneh, F., Thomas, J., & Wieland, L. S. (2019). Searching for and selecting studies. In *Cochrane handbook for systematic reviews of interventions* (pp. 67–107). Wiley.

Lockwood, O., & Si, M. (2020, October). Reinforcement learning with quantum variational circuit. In *Proceedings of the AAAI Conference on Artificial Intelligence and Interactive Digital Entertainment* (Vol. 16, No. 1, pp. 245–251).

Lockwood, O., & Si, M. (2021, July). Playing Atari with hybrid quantum-classical reinforcement learning. In *NeurIPS 2020 workshop on pre-registration in machine learning* (pp. 285–301). PMLR.

Martín-Guerrero, J. D., & Lamata, L. (2021). Reinforcement learning and physics. *Applied Sciences, 11*(18), 8589.

Martín-Guerrero, J. D., & Lamata, L. (2022). Quantum machine learning: A tutorial. *Neurocomputing, 470*, 457–461.

Melnikov, A., Kordzanganeh, M., Alodjants, A., & Lee, R. K. (2023). Quantum machine learning: From physics to software engineering. *Advances in Physics: X, 8*(1), 2165452.

Miguel-Ramiro, J., Pirker, A., & Dür, W. (2021). Genuine quantum networks with superposed tasks and addressing. *npj Quantum Information, 7*(1), 135.

Mishra, N., Kapil, M., Rakesh, H., Anand, A., Mishra, N., Warke, A., Sarkar, S., Dutta, S., Gupta, S., Prasad Dash, A., & Gharat, R. (2021). Quantum machine learning: A review and current status. In *Data management, analytics and innovation: Proceedings of ICDMAI 2020* (Vol. 2, pp. 101–145).

Montúfar, G. (2018). Restricted Boltzmann machines: Introduction and review. In *Information geometry and its applications: On the occasion of Shun-ichi Amari's 80th birthday, IGAIA IV Liblice, Czech Republic, June 2016* (pp. 75–115). Springer International Publishing.

Niraula, D., Jamaluddin, J., Matuszak, M. M., Haken, R. K. T., & Naqa, I. E. (2021). Quantum deep reinforcement learning for clinical decision support in oncology: Application to adaptive radiotherapy. *Scientific Reports, 11*(1), 23545.

Niu, M. Y., Boixo, S., Smelyanskiy, V. N., & Neven, H. (2019). Universal quantum control through deep reinforcement learning. *npj Quantum Information, 5*(1), 33.

Olivera-Atencio, M. L., Lamata, L., & Casado-Pascual, J. (2023). *Benefits of open quantum systems for quantum machine learning*. arXiv preprint arXiv:2308.02837

Plaat, A. (2022). *Deep reinforcement learning* (Vol. 10, pp. 978–981). Springer.

Prajapati, J. B., Paliwal, H., Prajapati, B. G., Saikia, S., & Pandey, R. (2023). Quantum machine learning in prediction of breast cancer. In *Quantum computing: A shift from bits to qubits* (pp. 351–382). Springer Nature Singapore.

Pramod, A., Naicker, H. S., & Tyagi, A. K. (2021). Machine learning and deep learning: Open issues and future research directions for the next 10 years. In *Computational analysis and deep learning for medical care: Principles, methods, and applications* (pp. 463–490). Wiley.

Rajawat, A. S., Goyal, S. B., Bedi, P., Jan, T., Whaiduzzaman, M., & Prasad, M. (2023). Quantum machine learning for security assessment in the internet of medical things (IoMT). *Future Internet, 15*(8), 271.

Saggio, V., Asenbeck, B. E., Hamann, A., Strömberg, T., Schiansky, P., Dunjko, V., Friis, N., Harris, N. C., Hochberg, M., Englund, D., & Wölk, S. (2021). Experimental quantum speed-up in reinforcement learning agents. *Nature, 591*(7849), 229–233.

Sajwan, P., & Jayapandian, N. (2019, December). Challenges and opportunities: Quantum computing in machine learning. In *2019 Third International Conference on I-SMAC (IoT in Social, Mobile, Analytics and Cloud)(I-SMAC)* (pp. 598–602). IEEE.

Sanchez-Palencia, L. (2018). *Quantum simulation: From basic principles to applications*. arXiv preprint arXiv:1812.01110

Sarkar, A., Al-Ars, Z., Gandhi, H., & Bertels, K. (2021). *QKSA: Quantum knowledge seeking agent—Resource-optimized reinforcement learning using quantum process tomography*. arXiv preprint arXiv:2112.03643

Schuld, M., & Petruccione, F. (2018). *Supervised learning with quantum computers* (Vol. 17). Springer.

Schuld, M., & Petruccione, F. (2021). *Machine learning with quantum computers*. Springer.

Sun, H., Zhang, Y., Baleanu, D., Chen, W., & Chen, Y. (2018). A new collection of real-world applications of fractional calculus in science and engineering. *Communications in Nonlinear Science and Numerical Simulation, 64*, 213–231.

Tacchino, F., Chiesa, A., Carretta, S., & Gerace, D. (2020). Quantum computers as universal quantum simulators: State-of-the-art and perspectives. *Advanced Quantum Technologies, 3*(3), 1900052.

Telahun, M. (2020). *Exploring information for quantum machine learning models*.

Wang, X., Lin, Z., Che, L., Chen, H., & Lu, D. (2022). Experimental quantum-enhanced machine learning in spin-based systems. *Advanced Quantum Technologies, 5*(8), 2200005.

Yun, W. J., Park, J., & Kim, J. (2023, June). Quantum multi-agent meta reinforcement learning. *Proceedings of the AAAI Conference on Artificial Intelligence, 37*(9), 11087–11095.

Zhou, D., Gu, Q., & Szepesvari, C. (2021, July). Nearly minimax optimal reinforcement learning for linear mixture Markov decision processes. In *Conference on learning theory* (pp. 4532–4576). PMLR.

The Legality of Space Cyber Security: "When Could Space Cyber Security Tend Towards Infringing Legality?"

Simon M. Wilson

Abstract The Space Domain is increasingly being exploited and there is no sign of this slowing. The fact that for you, your children or grandchildren, it is likely that there have been human beings in orbit above the Earth for you or their entire life is testimony to how we have tapped the resource that is near space. Just as in the terrestrial domain, cyber security is a vital consideration. The threat of cyber attack is very real and just as damaging. To enact an attack, in simple terms, first the target must be discovered; secondly its boundary must be breached; privilege attained; possible lateral movement; and an attack executed. Each stage requires protection, although potentially by protecting the previous stage. Numerous effects support making services cyber secure by the employment of various Cyber Resiliency Implementation Approaches and the subsequent security controls that are used to implement them. Where could the methods used to achieve the objectives of these cyber resiliency approaches and the subsequent effect breach the various national and international legal requirements across the world?
Example:

Effect:	Deceive
Resiliency Approach:	Disinformation
Candidate Mitigation:	Present Deceptive Information (CM1101)—Present deceptive information about systems, data, processes, and users. Monitor uses or search for presence of that information.
Potential Illegality:	Intentional misreporting of the geographical position that a space asset is in orbit above.

Although the above example may have legitimacy in a time of conflict between nations, during peacetime, this could likely be viewed as an illegal act. This chapter explores the possibility of cyber effects infringing law by looking at examples of where achieving cyber secure space services may potentially cause such an infringement, how this may be policed or enforced, or not, with the author's conclusion.

S. M. Wilson (✉)
Chester, UK
e-mail: ea.and.se@gmail.com

H. Jahankhani et al. (eds.), *Space Law Principles and Sustainable Measures*, Space Law and Policy, https://doi.org/10.1007/978-3-031-64045-2_5

Keywords Disinformation · Space Cyber Security · Space law · MITRE Adversarial Tactics · Cyber resilience

1 Introduction

This chapter is intended to be thought provoking. Are there limits to the ways we protect our assets? May we infringe legality in doing so and what are the ramifications? Indeed, considering the domain, how ultimately could these be enforced?

The Space Domain is increasingly being exploited and there is no sign of this slowing. The fact that for you, your children or grandchildren, it is likely that there have been human beings in orbit above the Earth for you or their entire life is testimony to how we have tapped the resource that is near space.
From Geostationary/Geospatial Orbit (GEO) and Geosynchronous Orbit (GSO), through Medium Earth Orbit (MEO), to the increasing use of Low Earth Orbit (LEO); non-terrestrial communications are highly exploited. In the days of high kinetic threat, moving these communications channels out of perceived reach provided higher physical resilience and minimised the footprint of vulnerable target assets. As such, some would say that the further away the better.

Satellite Orbit Altitudes (Heukelman, 2018)

Abbreviation	Orbit name	Altitude [km]
LEO	Low Earth Orbit	160 to 2000
MEO	Medium Earth Orbit	2000 to <35,786
GSO	Geosynchronous Orbit	35,786
GEO	Geostationary Equatorial Orbit	35,786

Today's world is different. We are still faced with the kinetic threat, but linked with a growing non-kinetic threat. This non-kinetic threat includes Cyber and Electro-Magnetic Activities (CEMA); as well as more soft skill threats, such as information operations, social profiling and exploitation. Whereas kinetic action tends to be overt, non-kinetic action tends to the opposite in being covert, although not exclusively. Coupled with a world that is becoming more and more politically fragmented and dangerous, the threat is only increasing. Distance no longer provides resilience.

We have been living with this non-kinetic Cyber and Electro-Magnetic Activity threat for some time now, but the reliance and increased use of the Space Domain means we must be evermore focused on its protection; especially in light of increased Offensive Cyber (OC) capabilities and intent.

2 Cyber Attack

For an adversary to execute an attack within Cyberspace, they first of all need to be able to discover the target—the MITRE Adversarial Tactics, Techniques, & Common Knowledge (ATT&CK) framework (MITRE, n.d.) calls this the Reconnaissance 'Tactic'.

Secondly they will need to be able to breach the boundary of the target, which MITRE covers with the Resource Development and Initial Access tactics.

Onwards from this is establishing and escalating privilege; lateral movement; discovering the target resource; and then exploiting and/or attacking the target asset. Throughout all of this activity the adversary will be trying to evade detection. MITRE cover all of these steps within the ATT&CK framework.

We obviously wish to protect against such activity. We must though, be conscious of the legality of our own actions. Can we be guilty of illegal actions in trying to protect our own assets?

Numerous effects support making services cyber secure by the employment of various Cyber Resiliency Implementation Approaches and the subsequent security controls that are used to implement them.

The US National Institute of Standards and Technology (NIST) has defined 15 Cyber Effects to contribute towards achieving system cyber resiliency (NIST, 2021). These are listed in the table below with a short description; and fuller details that includes lists of associated cyber resiliency implementation approaches included in the next section, which it is hoped proves to be good reference material for the reader.

Cyber effects (NIST, 2021)

Cyber effect	Description
Contain	Restrict the effects of the threat event to a limited set of resources
Deceive	Lead the adversary to believe false information about individuals, systems, missions, organisations, defender capabilities, or TTPs
Degrade	Decrease the expected consequences of the threat event
Delay	Increase the amount of time needed for the threat event to result in adverse impacts
Detect	Identify a threat event or its effects by discovering or discerning the fact that the event is occurring, has occurred, or (based on indicators, warnings, and precursor activities) is about to occur
Deter	Discourage the adversary from taking an action by instilling fear (e.g., of attribution or retribution) or doubt that the action would achieve intended effects (e.g., that targets exist)
Divert	Direct the threat event toward or away from defender-chosen resources
Exert	Increase the level of effort or resources needed for an adversary to achieve a given result
Expunge	Remove resources that are known to be or suspected of being unsafe, incorrect, or corrupted
Negate	Create conditions under which the threat event cannot be expected to result in an impact

(continued)

(continued)

Cyber effect	Description
Pre-empt	Forestall or avoid conditions under which the threat event could occur
Reduce	Decrease the degree of damage from a threat event. The degree of damage can have two dimensions: breadth (i.e., number of affected resources) and depth (i.e., level of harm to a given resource)
Reveal	Share information about risk factors and the relative effectiveness of remediation approaches with partners, stakeholder community, or the general public
Scrutinise	Analyse threat events and artifacts associated with threat events to develop indicators, determine sources of events, assess damage, and identify patterns of exploiting vulnerabilities, predisposing conditions, and weaknesses
Shorten	Limit the duration of adverse consequences of a threat event

3 Cyber Effect to Cyber Resiliency Implementation Approach (NIST, 2021)

This section provides details derived from, but not provided by, the US National Institute of Standards and Technology (NIST) in the Special Publication (SP) NIST SP 800-160v2r1 (NIST, 2021).

For each of the 15 Cyber Effects that NIST identify (table above), the associated Cyber Resiliency Implementation Approaches have been listed with a short description, as provided by NIST. These have been reverse engineered from the mappings that NIST provide in their analysis, starting from the MITRE Adversarial Tactics, Techniques, & Common Knowledge (ATT&CK) framework (MITRE, n.d.).

The following sub-sections provide a brief description of each cyber effect followed by a table of its associated cyber resiliency implementation approaches.

4 Contain

Restrict the effects of the threat event to a limited set of resources.

The consequences of the threat event are less extensive than they might otherwise be.

Reduce the level of impact.

- The adversary can affect fewer resources than planned. The value of the activity to the adversary, in terms of achieving the adversary's goals, is reduced.

Example: The defender organisation makes changes to a combination of internal firewalls and logically separated networks (dynamic segmentation) to isolate enclaves in response to the detection of malware, limiting the effects of the malware to initially infected enclaves.

Resiliency approach	Description
Dynamic reconfiguration	Make changes to individual systems, system elements, components, or sets of cyber resources to change functionality or behaviour without interrupting service
Adaptive management	Change how mechanisms are used based on changes in the operational environment as well as changes in the threat environment
Misdirection	Maintain deception resources or environments and direct adversary activities there
Dynamic privileges	Elevate or decrease privileges assigned to a user, process, or service based on transient or contextual factors
Restriction	Remove or disable unneeded functionality or connectivity, or add mechanisms to reduce the chance of vulnerability or failure
Predefined segmentation	Define enclaves, segments, or other types of resource sets based on criticality and trustworthiness so that they can be protected separately and, if necessary, isolated
Dynamic segmentation and isolation	Change the configuration of enclaves or protected segments, or isolate resources while minimising operational disruption
Integrity checks	Apply and validate checks of the integrity or quality of information, components, or services

5 Deceive

Lead the adversary to believe false information about individuals, systems, missions, organisations, defender capabilities, or TTPs.

This effect is relevant only to adversarial threat events and involves influencing the adversary's actions.

Reduce the likelihood of occurrence, and/or reduce the likelihood of impact.

- The adversary's efforts are wasted as the assumptions on which the adversary bases attacks are false.
- The adversary takes actions based on false information, thus revealing that they have obtained that information.

Example: The defender strategically places false information (disinformation) about the cybersecurity investments that it plans to make. As a result, the adversary's malware development is wasted by being focused on countering non-existent cybersecurity protections.

Example: The defender uses selectively planted false information (disinformation) and honeynets (misdirection) to cause an adversary to focus its malware on virtual sandboxes while simultaneously employing obfuscation to hide the actual resources.

Resiliency approach	Description
Disinformation	Provide deliberately misleading information to adversaries
Misdirection	Maintain deception resources or environments and direct adversary activities there
Tainting	Embed covert capabilities in resources
Architectural diversity	Use multiple sets of technical standards, different technologies, and different architectural patterns

6 Degrade

Decrease the expected consequences of the threat event.

Because the consequences of the threat event are less severe than they would be without the mitigation, they could fail to produce an impact, or their impact could be lessened.

Reduce the likelihood of impact, and/or reduce the level of impact.

- Not all of the resources targeted by the adversary are affected, or the targeted resources are affected to a lesser degree than the adversary sought.

Example: The defender uses multiple browsers and operating systems (architectural diversity) on both end-user systems and some critical servers. The result is that malware targeted at specific software can only compromise a subset of the targeted systems; a sufficient number continue to operate to complete the mission or business function.

Resiliency approach	Description
Dynamic reconfiguration	Make changes to individual systems, system elements, components, or sets of cyber resources to change functionality or behaviour without interrupting service
Dynamic resource allocation	Change the allocation of resources to tasks or functions without terminating critical functions or processes
Adaptive management	Change how mechanisms are used based on changes in the operational environment as well as changes in the threat environment
Dynamic threat awareness	Maintain current information about threat actors, indicators, and potential, predicted, and observed adverse events
Calibrated defence-in-depth	Provide complementary protective mechanisms at different architectural layers or in different locations, calibrating the strength and number of mechanisms to resource value

(continued)

(continued)

Resiliency approach	Description
Consistency analysis	Determine whether and how protections can be applied in a coordinated, consistent way that minimises interference, potential cascading failures, or coverage gaps
Obfuscation	Hide, transform, or otherwise obfuscate information from the adversary
Disinformation	Provide deliberately misleading information to adversaries
Misdirection	Maintain deception resources or environments and direct adversary activities there
Architectural diversity	Use multiple sets of technical standards, different technologies, and different architectural patterns
Path diversity	Provide multiple independent paths for command, control, and communications
Functional relocation cyber resources	Change the location of cyber resources that provide functionality or information, either by moving the assets or by transferring functional responsibility
Non-persistent information	Refresh information periodically, or generate information on demand, and delete it when no longer needed
Non-persistent connectivity	Establish connections on demand, and terminate connections when no longer needed
Trust based privilege management	Define, assign, and maintain privileges associated with active entities based on established trust criteria consistent with principles of least privilege
Attribute-based usage restrictions	Define, assign, maintain, and apply usage restrictions on systems containing cyber resources based on the criticality of missions or business functions and other attributes (e.g., data sensitivity)
Dynamic privileges	Elevate or decrease privileges assigned to a user, process, or service based on transient or contextual factors
Purposing	Ensure systems containing cyber resources are used consistently with mission or business function purposes and approved uses
Restriction	Remove or disable unneeded functionality or connectivity, or add mechanisms to reduce the chance of vulnerability or failure
Replication	Duplicate hardware, information, backups, or functionality in multiple locations and keep them synchronised
Predefined segmentation	Define enclaves, segments, or other types of resource sets based on criticality and trustworthiness so that they can be protected separately and, if necessary, isolated
Dynamic segmentation and isolation	Change the configuration of enclaves or protected segments, or isolate resources while minimising operational disruption
Integrity checks	Apply and validate checks of the integrity or quality of information, components, or services
Provenance tracking	Identify and track the provenance of data, software, or hardware elements

(continued)

(continued)

Resiliency approach	Description
Temporal unpredictability	Change behaviour or state at times that are determined randomly or by complex functions

7 Delay

Increase the amount of time needed for the threat event to result in adverse impacts.

Because the consequences of the threat event occur later than they would without the mitigation, they could fail to produce an impact, or their impact could be lessened.

Reduce the likelihood of impact, and/or reduce the level of impact.

- The adversary achieves the intended effects but not within the intended period.

Example: The protection measures (e.g., access controls, encryption) allocated to resources increase in number and strength based on resource criticality (calibrated defence-in-depth). The frequency of authentication challenges varies randomly (temporal unpredictability) and more often for more critical resources. The result is that it takes the attacker more time to successfully compromise the targeted resources.

Resiliency approach	Description
Dynamic reconfiguration	Make changes to individual systems, system elements, components, or sets of cyber resources to change functionality or behaviour without interrupting service
Adaptive management	Change how mechanisms are used based on changes in the operational environment as well as changes in the threat environment
Calibrated defence-in-depth	Provide complementary protective mechanisms at different architectural layers or in different locations, calibrating the strength and number of mechanisms to resource value
Consistency analysis	Determine whether and how protections can be applied in a coordinated, consistent way that minimises interference, potential cascading failures, or coverage gaps
Obfuscation	Hide, transform, or otherwise obfuscate information from the adversary
Disinformation	Provide deliberately misleading information to adversaries
Misdirection	Maintain deception resources or environments and direct adversary activities there
Architectural diversity	Use multiple sets of technical standards, different technologies, and different architectural patterns
Design diversity	Use different designs to meet the same requirements or provide equivalent functionality
Synthetic diversity	Transform implementations of software to produce a variety of instances

(continued)

(continued)

Resiliency approach	Description
Path diversity	Provide multiple independent paths for command, control, and communications
Functional relocation cyber resources	Change the location of cyber resources that provide functionality or information, either by moving the assets or by transferring functional responsibility
Fragmentation	Fragment information and distribute it across multiple components
Non-persistent information	Refresh information periodically, or generate information on demand, and delete it when no longer needed
Non-persistent connectivity	Establish connections on demand, and terminate connections when no longer needed
Trust based privilege management	Define, assign, and maintain privileges associated with active entities based on established trust criteria consistent with principles of least privilege
Attribute-based usage restrictions	Define, assign, maintain, and apply usage restrictions on systems containing cyber resources based on the criticality of missions or business functions and other attributes (e.g., data sensitivity)
Dynamic privileges	Elevate or decrease privileges assigned to a user, process, or service based on transient or contextual factors
Purposing	Ensure systems containing cyber resources are used consistently with mission or business function purposes and approved uses
Restriction	Remove or disable unneeded functionality or connectivity, or add mechanisms to reduce the chance of vulnerability or failure
Predefined segmentation	Define enclaves, segments, or other types of resource sets based on criticality and trustworthiness so that they can be protected separately and, if necessary, isolated
Dynamic segmentation and isolation	Change the configuration of enclaves or protected segments, or isolate resources while minimising operational disruption
Integrity checks	Apply and validate checks of the integrity or quality of information, components, or services
Provenance tracking	Identify and track the provenance of data, software, or hardware elements
Temporal unpredictability	Change behaviour or state at times that are determined randomly or by complex functions
Contextual unpredictability	Change behaviour or state in ways that are determined randomly or by complex functions

8 Detect

Identify a threat event or its effects by discovering or discerning the fact that the event is occurring, has occurred, or (based on indicators, warnings, and precursor activities) is about to occur.

Detection informs corrective actions.

Reduce the likelihood of impact, and reduce the level of impact (depending on responses).

- The adversary's activities become susceptible to defensive responses.

Example: The defender continually moves its sensors (functional relocation of sensors), often at random times (temporal unpredictability), to common points of egress from the organisation. They combine this with the use of beacon traps (tainting). The result is that the defender can quickly detect efforts by the adversary to exfiltrate sensitive information.

Resiliency approach	Description
Dynamic reconfiguration	Make changes to individual systems, system elements, components, or sets of cyber resources to change functionality or behaviour without interrupting service
Dynamic resource allocation	Change the allocation of resources to tasks or functions without terminating critical functions or processes
Adaptive management	Change how mechanisms are used based on changes in the operational environment as well as changes in the threat environment
Monitoring and damage assessment	Monitor and analyse behaviour and characteristics of components and resources to look for indicators of adversary activity and to detect and assess damage from adversity
Sensor fusion and analysis	Fuse and analyse monitoring data and analysis results from different information sources or at different times together with externally provided threat intelligence
Forensic and behavioural analysis	Analyse adversary Tactics, Techniques and Procedures (TTP), including observed behaviour as well as malware and other artefacts left behind by adverse events
Dynamic resource awareness	Maintain current information about resources, status of resources, and resource connectivity
Dynamic threat awareness	Maintain current information about threat actors, indicators, and potential, predicted, and observed adverse events
Mission dependency and status visualisation	Maintain current information about the status of missions or business functions, dependencies on resources, and the status of those resources with respect to threats
Consistency analysis	Determine whether and how protections can be applied in a coordinated, consistent way that minimises interference, potential cascading failures, or coverage gaps

(continued)

(continued)

Resiliency approach	Description
Self-challenge	Affect mission/business processes or system elements adversely in a controlled manner to validate the effectiveness of protections and to enable proactive response and improvement
Disinformation	Provide deliberately misleading information to adversaries
Misdirection	Maintain deception resources or environments and direct adversary activities there
Tainting	Embed covert capabilities in resources
Architectural diversity	Use multiple sets of technical standards, different technologies, and different architectural patterns
Path diversity	Provide multiple independent paths for command, control, and communications
Functional relocation of sensors	Relocate sensors or reallocate responsibility for specific sensing tasks to look for indicators of adverse events
Integrity checks	Apply and validate checks of the integrity or quality of information, components, or services
Provenance tracking	Identify and track the provenance of data, software, or hardware elements
Behaviour validation	Validate the behaviour of a system, service, or device against defined or emergent criteria (e.g., requirements, patterns of prior usage)

9 Deter

Discourage the adversary from taking an action by instilling fear (e.g., of attribution or retribution) or doubt that the action would achieve intended effects (e.g., that targets exist).

This effect is relevant only to adversarial threat events and involves influencing the adversary's decision-making process.

Reduce the likelihood of occurrence.

- The adversary ceases or suspends activities.

Example: The defender uses disinformation to make it appear that the organisation is better able to detect attacks than it is and is willing to launch major counter-strikes. Therefore, the adversary chooses to not launch an attack due to fear of detection and reprisal.

Resiliency approach	Description
Disinformation	Provide deliberately misleading information to adversaries

10 Divert

Direct the threat event toward defender-chosen resources.

The event affects resources that the defender does not care about or for which the defender can manage consequences.

Reduce the likelihood of occurrence.

- The adversary refocuses activities on defender-chosen resources.
- The adversary directs activities toward targets beyond the defender's purview (e.g., other organisations).
- The adversary does not affect resources that the defender has not selected to be targets.

Example: The defender maintains an Internet-visible enclave with which untrusted external entities can interact and a private enclave accessible only via a VPN for trusted suppliers, partners, or customers (predefined segmentation).

Example: The defender uses non-persistent information and obfuscation to hide critical resources combined with functional relocation of cyber resources and disinformation to lure the adversary toward a sandboxed enclave where adversary actions cannot harm critical resources.

Resiliency approach	Description
Misdirection	Maintain deception resources or environments and direct adversary activities there
Architectural diversity	Use multiple sets of technical standards, different technologies, and different architectural patterns
Predefined segmentation	Define enclaves, segments, or other types of resource sets based on criticality and trustworthiness so that they can be protected separately and, if necessary, isolated

11 Exert

Increase the level of effort or resources needed for an adversary to achieve a given result.

This effect is relevant only to adversarial threat events and involves increasing the adversary's costs.

Reduce the likelihood of impact.

- The adversary gives up planned or partially completed activities in response to finding that additional effort or resources are needed.
- The adversary achieves the intended effects in their desired time frame but only by applying more resources. Thus, the adversary's return on investment (ROI) is decreased.
- The adversary reveals TTPs they had planned to reserve for future use.

Example: The defender enhances the defences of moderate-criticality components with additional mitigations (calibrated defence-in-depth). To overcome these, the adversary must tailor and deploy TTPs that they were planning to reserve for use against higher value defender targets.

Example: The defender adds a large amount of valid but useless information to a data store (obfuscation), requiring the adversary to exfiltrate and analyse more data before taking further actions.

Resiliency approach	Description
Dynamic reconfiguration	Make changes to individual systems, system elements, components, or sets of cyber resources to change functionality or behaviour without interrupting service
Adaptive management	Change how mechanisms are used based on changes in the operational environment as well as changes in the threat environment
Dynamic threat awareness	Maintain current information about threat actors, indicators, and potential, predicted, and observed adverse events
Calibrated defence-in-depth	Provide complementary protective mechanisms at different architectural layers or in different locations, calibrating the strength and number of mechanisms to resource value
Consistency analysis	Determine whether and how protections can be applied in a coordinated, consistent way that minimises interference, potential cascading failures, or coverage gaps
Orchestration	Coordinate the ongoing behaviour of mechanisms and processes at different layers, in different locations, or implemented for different aspects of trustworthiness to avoid causing cascading failures, interference, or coverage gaps
Self-challenge	Affect mission/business processes or system elements adversely in a controlled manner to validate the effectiveness of protections and to enable proactive response and improvement
Obfuscation	Hide, transform, or otherwise obfuscate information from the adversary
Disinformation	Provide deliberately misleading information to adversaries
Misdirection	Maintain deception resources or environments and direct adversary activities there
Tainting	Embed covert capabilities in resources
Architectural diversity	Use multiple sets of technical standards, different technologies, and different architectural patterns
Design diversity	Use different designs to meet the same requirements or provide equivalent functionality

(continued)

(continued)

Resiliency approach	Description
Synthetic diversity	Transform implementations of software to produce a variety of instances
Information diversity	Provide information from different sources or transform information in different ways
Path diversity	Provide multiple independent paths for command, control, and communications
Supply chain diversity	Use multiple independent supply chains for critical components
Functional relocation cyber resources	Change the location of cyber resources that provide functionality or information, either by moving the assets or by transferring functional responsibility
Asset mobility	Securely move physical resources
Fragmentation	Fragment information and distribute it across multiple components
Distributed functionality	Decompose a function or application into smaller functions and distribute those functions across multiple components
Non-persistent information	Refresh information periodically, or generate information on demand, and delete it when no longer needed
Non-persistent connectivity	Establish connections on demand, and terminate connections when no longer needed
Trust based privilege management	Define, assign, and maintain privileges associated with active entities based on established trust criteria consistent with principles of least privilege
Attribute-based usage restrictions	Define, assign, maintain, and apply usage restrictions on systems containing cyber resources based on the criticality of missions or business functions and other attributes (e.g., data sensitivity)
Dynamic privileges	Elevate or decrease privileges assigned to a user, process, or service based on transient or contextual factors
Purposing	Ensure systems containing cyber resources are used consistently with mission or business function purposes and approved uses
Restriction	Remove or disable unneeded functionality or connectivity, or add mechanisms to reduce the chance of vulnerability or failure
Specialisation	Modify the design of, augment, or configure critical cyber resources uniquely for the mission or business function to improve trustworthiness
Protected backup and restore	Back up information and software (including configuration data and virtualised resources) in a way that protects its confidentiality, integrity, and authenticity, and enable restoration in case of disruption or corruption
Replication	Duplicate hardware, information, backups, or functionality in multiple locations and keep them synchronised
Predefined segmentation	Define enclaves, segments, or other types of resource sets based on criticality and trustworthiness so that they can be protected separately and, if necessary, isolated

(continued)

(continued)

Resiliency approach	Description
Dynamic segmentation and isolation	Change the configuration of enclaves or protected segments, or isolate resources while minimising operational disruption
Integrity checks	Apply and validate checks of the integrity or quality of information, components, or services
Provenance tracking	Identify and track the provenance of data, software, or hardware elements
Behaviour validation	Validate the behaviour of a system, service, or device against defined or emergent criteria (e.g., requirements, patterns of prior usage)
Temporal unpredictability	Change behaviour or state at times that are determined randomly or by complex functions
Contextual unpredictability	Change behaviour or state in ways that are determined randomly or by complex functions

12 Expunge

Remove resources that are known to be or suspected of being unsafe, incorrect, or corrupted.

Reduce the likelihood of impact of subsequent events in the same threat scenario.

A malfunctioning, misbehaving, or suspect resource is restored to normal operation.

- The adversary loses a capability for some period, as adversary-directed threat mechanisms (e.g., malicious code) are removed.
- Adversary-controlled resources are so badly damaged that they cannot perform any function or be restored to a usable condition without being entirely rebuilt.

Example: The defender uses virtualisation to refresh critical software (non-persistent services) from a known good copy at random intervals (temporal unpredictability). As a result, malware that was implanted in the software is deleted.

Resiliency approach	Description
Dynamic reconfiguration	Make changes to individual systems, system elements, components, or sets of cyber resources to change functionality or behaviour without interrupting service
Asset mobility	Securely move physical resources
Non-persistent information	Refresh information periodically, or generate information on demand, and delete it when no longer needed
Non-persistent services	Refresh services periodically, or generate services on demand and terminate services when no longer needed

(continued)

(continued)

Resiliency approach	Description
Provenance tracking	Identify and track the provenance of data, software, or hardware elements

13 Negate

Create conditions under which the threat event cannot be expected to result in an impact.

The threat event may produce consequences, but those consequences cannot produce an impact.

Reduce the likelihood of impact.

- The adversary can launch an attack, but it will not even partially succeed. The adversary's efforts are wasted as the assumptions on which the adversary based its attack are no longer valid, and as a result, the intended effects cannot be achieved.

Example: Subtle variations in critical software are implemented (synthetic diversity) and prevent the adversary's malware from compromising the targeted software.

Resiliency approach	Description
Monitoring and damage assessment	Monitor and analyse behaviour and characteristics of components and resources to look for indicators of adversary activity and to detect and assess damage from adversity
Dynamic threat awareness	Maintain current information about threat actors, indicators, and potential, predicted, and observed adverse events
Calibrated defence-in-depth	Provide complementary protective mechanisms at different architectural layers or in different locations, calibrating the strength and number of mechanisms to resource value
Consistency analysis	Determine whether and how protections can be applied in a coordinated, consistent way that minimises interference, potential cascading failures, or coverage gaps
Self-challenge	Affect mission/business processes or system elements adversely in a controlled manner to validate the effectiveness of protections and to enable proactive response and improvement
Obfuscation	Hide, transform, or otherwise obfuscate information from the adversary
Disinformation	Provide deliberately misleading information to adversaries
Misdirection	Maintain deception resources or environments and direct adversary activities there
Architectural diversity	Use multiple sets of technical standards, different technologies, and different architectural patterns

(continued)

(continued)

Resiliency approach	Description
Design diversity	Use different designs to meet the same requirements or provide equivalent functionality
Synthetic diversity	Transform implementations of software to produce a variety of instances
Path diversity	Provide multiple independent paths for command, control, and communications
Functional relocation cyber resources	Change the location of cyber resources that provide functionality or information, either by moving the assets or by transferring functional responsibility
Non-persistent information	Refresh information periodically, or generate information on demand, and delete it when no longer needed
Non-persistent services	Refresh services periodically, or generate services on demand and terminate services when no longer needed
Trust based privilege management	Define, assign, and maintain privileges associated with active entities based on established trust criteria consistent with principles of least privilege
Attribute-based usage restrictions	Define, assign, maintain, and apply usage restrictions on systems containing cyber resources based on the criticality of missions or business functions and other attributes (e.g., data sensitivity)
Purposing	Ensure systems containing cyber resources are used consistently with mission or business function purposes and approved uses
Restriction	Remove or disable unneeded functionality or connectivity, or add mechanisms to reduce the chance of vulnerability or failure
Replication	Duplicate hardware, information, backups, or functionality in multiple locations and keep them synchronised
Predefined segmentation	Define enclaves, segments, or other types of resource sets based on criticality and trustworthiness so that they can be protected separately and, if necessary, isolated
Dynamic segmentation and isolation	Change the configuration of enclaves or protected segments, or isolate resources while minimising operational disruption
Integrity checks	Apply and validate checks of the integrity or quality of information, components, or services
Provenance tracking	Identify and track the provenance of data, software, or hardware elements
Behaviour validation	Validate the behaviour of a system, service, or device against defined or emergent criteria (e.g., requirements, patterns of prior usage)
Temporal unpredictability	Change behaviour or state at times that are determined randomly or by complex functions

14 Pre-empt

Forestall or avoid conditions under which the threat event could occur.

The threat event cannot have any consequences because it cannot actually occur.

Reduce the likelihood of occurrence.

- The adversary's resources cannot be applied, or the adversary cannot perform activities (e.g., because resources adversary requires are destroyed or made inaccessible).

Example: An unneeded network connection is disabled (non-persistent connectivity) so that an attack via that interface cannot be made.

Example: A resource is repositioned (asset mobility) so that it cannot be affected by a threat event in its new location.

Resiliency approach	Description
Dynamic reconfiguration	Make changes to individual systems, system elements, components, or sets of cyber resources to change functionality or behaviour without interrupting service
Adaptive management	Change how mechanisms are used based on changes in the operational environment as well as changes in the threat environment
Dynamic threat awareness	Maintain current information about threat actors, indicators, and potential, predicted, and observed adverse events
Calibrated defence-in-depth	Provide complementary protective mechanisms at different architectural layers or in different locations, calibrating the strength and number of mechanisms to resource value
Consistency analysis	Determine whether and how protections can be applied in a coordinated, consistent way that minimises interference, potential cascading failures, or coverage gaps
Self-challenge	Affect mission/business processes or system elements adversely in a controlled manner to validate the effectiveness of protections and to enable proactive response and improvement
Obfuscation	Hide, transform, or otherwise obfuscate information from the adversary
Disinformation	Provide deliberately misleading information to adversaries
Architectural diversity	Use multiple sets of technical standards, different technologies, and different architectural patterns
Synthetic diversity	Transform implementations of software to produce a variety of instances
Functional relocation of sensors	Relocate sensors or reallocate responsibility for specific sensing tasks to look for indicators of adverse events

(continued)

(continued)

Resiliency approach	Description
Functional relocation cyber resources	Change the location of cyber resources that provide functionality or information, either by moving the assets or by transferring functional responsibility
Non-persistent information	Refresh information periodically, or generate information on demand, and delete it when no longer needed
Non-persistent services	Refresh services periodically, or generate services on demand and terminate services when no longer needed
Non-persistent connectivity	Establish connections on demand, and terminate connections when no longer needed
Trust based privilege management	Define, assign, and maintain privileges associated with active entities based on established trust criteria consistent with principles of least privilege
Purposing	Ensure systems containing cyber resources are used consistently with mission or business function purposes and approved uses
Restriction	Remove or disable unneeded functionality or connectivity, or add mechanisms to reduce the chance of vulnerability or failure
Predefined segmentation	Define enclaves, segments, or other types of resource sets based on criticality and trustworthiness so that they can be protected separately and, if necessary, isolated
Dynamic segmentation and isolation	Change the configuration of enclaves or protected segments, or isolate resources while minimising operational disruption
Integrity checks	Apply and validate checks of the integrity or quality of information, components, or services
Provenance tracking	Identify and track the provenance of data, software, or hardware elements
Temporal unpredictability	Change behaviour or state at times that are determined randomly or by complex functions

15 Reduce

Decrease the degree of damage from a threat event.

The degree of damage can have two dimensions: breadth (i.e., number of affected resources) and depth (i.e., level of harm to a given resource).

A decrease in the degree of damage lessens the impact.

Reduce the level of impact.

- The level of damage to organisational missions or business operations from adversary activities is reduced due to partial restoration or reconstitution of all affected resources.

Example: Resources determined to be corrupted or suspect (integrity checks, behaviour validation) are restored from older, uncorrupted resources (protected backup and restore) with reduced functionality.

- The level of damage to organisational missions or business operations from adversary activities is reduced due to full restoration or reconstitution of some of the affected resources.

Example: The organisation removes one of three compromised resources and provides a new resource (replacement, specialisation) for the same or equivalent mission or business functionality.

Resiliency approach	Description
Dynamic reconfiguration	Make changes to individual systems, system elements, components, or sets of cyber resources to change functionality or behaviour without interrupting service
Dynamic resource allocation	Change the allocation of resources to tasks or functions without terminating critical functions or processes
Adaptive management	Change how mechanisms are used based on changes in the operational environment as well as changes in the threat environment
Consistency analysis	Determine whether and how protections can be applied in a coordinated, consistent way that minimises interference, potential cascading failures, or coverage gaps
Orchestration	Coordinate the ongoing behaviour of mechanisms and processes at different layers, in different locations, or implemented for different aspects of trustworthiness to avoid causing cascading failures, interference, or coverage gaps
Self-challenge	Affect mission/business processes or system elements adversely in a controlled manner to validate the effectiveness of protections and to enable proactive response and improvement
Architectural diversity	Use multiple sets of technical standards, different technologies, and different architectural patterns
Design diversity	Use different designs to meet the same requirements or provide equivalent functionality
Information diversity	Provide information from different sources or transform information in different ways
Path diversity	Provide multiple independent paths for command, control, and communications
Fragmentation	Fragment information and distribute it across multiple components
Non-persistent connectivity	Establish connections on demand, and terminate connections when no longer needed

(continued)

(continued)

Resiliency approach	Description
Trust based privilege management	Define, assign, and maintain privileges associated with active entities based on established trust criteria consistent with principles of least privilege
Dynamic privileges	Elevate or decrease privileges assigned to a user, process, or service based on transient or contextual factors
Specialisation	Modify the design of, augment, or configure critical cyber resources uniquely for the mission or business function to improve trustworthiness
Protected backup and restore	Back up information and software (including configuration data and virtualised resources) in a way that protects its confidentiality, integrity, and authenticity, and enable restoration in case of disruption or corruption
Surplus capacity	Maintain extra capacity for information storage, processing, or communications
Replication	Duplicate hardware, information, backups, or functionality in multiple locations and keep them synchronised
Predefined segmentation	Define enclaves, segments, or other types of resource sets based on criticality and trustworthiness so that they can be protected separately and, if necessary, isolated
Dynamic segmentation and isolation	Change the configuration of enclaves or protected segments, or isolate resources while minimising operational disruption
Integrity checks	Apply and validate checks of the integrity or quality of information, components, or services
Provenance tracking	Identify and track the provenance of data, software, or hardware elements

16 Reveal

Share information about risk factors and the relative effectiveness of remediation approaches with partners, stakeholder community, or the general public.

Threat information sharing supports common, joint, or coordinated risk responses. Information about threat events can be shared broadly or with a limited set of threat intelligence information-sharing partners.

Reduce the likelihood of impact, particularly in the future.

- The adversary loses the advantage of surprise and plausible deniability.
- The adversary's ability to compromise one organisation's systems to attack another organisation is impaired as awareness of adversary characteristics and behaviour across the stakeholder community (e.g., across all computer security incident response teams that support a given sector that might be expected to be attacked by the same actor or actors) is increased.

Example: The defender participates in threat information sharing and uses dynamically updated threat intelligence data feeds (dynamic threat modelling) to inform actions (adaptive management).

Resiliency approach	Description
Monitoring and damage assessment	Monitor and analyse behaviour and characteristics of components and resources to look for indicators of adversary activity and to detect and assess damage from adversity
Dynamic threat awareness	Maintain current information about threat actors, indicators, and potential, predicted, and observed adverse events
Tainting	Embed covert capabilities in resources
Integrity checks	Apply and validate checks of the integrity or quality of information, components, or services

17 Scrutinise

Analyse threat events and artifacts associated with threat events to develop indicators, determine sources of events, assess damage, and identify patterns of exploiting vulnerabilities, predisposing conditions, and weaknesses.

Scrutiny informs more effective detection and risk response.

Reduce the likelihood of impact.

- The adversary loses the advantages of uncertainty, confusion, and doubt.
- The defender has a better understanding the adversary, based on an analysis of the adversary's activities, including the artifacts (e.g., malicious code) and effects associated with those activities and the correlation of activity-specific observations with other activities (as feasible), and, thus, can recognise adversary Tactics, Techniques and Procedures (TTP).

Example: The defender deploys honeynets (misdirection), inviting attacks by the adversary and allowing the adversary to apply its TTPs in a safe environment. The defender then analyses (malware and forensic analysis) the malware captured in the honeynet to determine the nature of the attacker's TTPs, allowing it to develop appropriate defences.

Resiliency approach	Description
Adaptive management	Change how mechanisms are used based on changes in the operational environment as well as changes in the threat environment

(continued)

(continued)

Resiliency approach	Description
Monitoring and damage assessment	Monitor and analyse behaviour and characteristics of components and resources to look for indicators of adversary activity and to detect and assess damage from adversity
Sensor fusion and analysis	Fuse and analyse monitoring data and analysis results from different information sources or at different times together with externally provided threat intelligence
Forensic and behavioural analysis	Analyse adversary Tactics, Techniques and Procedures (TTP), including observed behaviour as well as malware and other artefacts left behind by adverse events
Dynamic threat awareness	Maintain current information about threat actors, indicators, and potential, predicted, and observed adverse events
Mission dependency and status visualisation	Maintain current information about the status of missions or business functions, dependencies on resources, and the status of those resources with respect to threats
Self-challenge	Affect mission/business processes or system elements adversely in a controlled manner to validate the effectiveness of protections and to enable proactive response and improvement
Misdirection	Maintain deception resources or environments and direct adversary activities there
Tainting	Embed covert capabilities in resources
Information diversity	Provide information from different sources or transform information in different ways
Integrity checks	Apply and validate checks of the integrity or quality of information, components, or services
Provenance tracking	Identify and track the provenance of data, software, or hardware elements

18 Shorten

Limit the duration of adverse consequences of a threat event.

Because the consequences of the threat event do not persist as long as they would without the mitigation, they could fail to produce an impact, or their impact could be lessened.

Reduce the level of impact

- The time-period during which the adversary's activities affect defender resources is limited.

Example: The defender employs a diverse set of suppliers (supply chain diversity) for time-critical components. As a result, when an adversary's attack on one supplier causes it to shut down, the defender can increase its use of the other suppliers, thus shortening the time during which it is without the critical components.

Resiliency approach	Description
Dynamic reconfiguration	Make changes to individual systems, system elements, components, or sets of cyber resources to change functionality or behaviour without interrupting service
Dynamic resource allocation	Change the allocation of resources to tasks or functions without terminating critical functions or processes
Adaptive management	Change how mechanisms are used based on changes in the operational environment as well as changes in the threat environment
Monitoring and damage assessment	Monitor and analyse behaviour and characteristics of components and resources to look for indicators of adversary activity and to detect and assess damage from adversity
Consistency analysis	Determine whether and how protections can be applied in a coordinated, consistent way that minimises interference, potential cascading failures, or coverage gaps
Orchestration	Coordinate the ongoing behaviour of mechanisms and processes at different layers, in different locations, or implemented for different aspects of trustworthiness to avoid causing cascading failures, interference, or coverage gaps
Self-challenge	Affect mission/business processes or system elements adversely in a controlled manner to validate the effectiveness of protections and to enable proactive response and improvement
Obfuscation	Hide, transform, or otherwise obfuscate information from the adversary
Architectural diversity	Use multiple sets of technical standards, different technologies, and different architectural patterns
Design diversity	Use different designs to meet the same requirements or provide equivalent functionality
Information diversity	Provide information from different sources or transform information in different ways
Path diversity	Provide multiple independent paths for command, control, and communications
Functional relocation cyber resources	Change the location of cyber resources that provide functionality or information, either by moving the assets or by transferring functional responsibility
Fragmentation	Fragment information and distribute it across multiple components
Distributed functionality	Decompose a function or application into smaller functions and distribute those functions across multiple components

(continued)

(continued)

Resiliency approach	Description
Non-persistent information	Refresh information periodically, or generate information on demand, and delete it when no longer needed
Non-persistent services	Refresh services periodically, or generate services on demand and terminate services when no longer needed
Non-persistent connectivity	Establish connections on demand, and terminate connections when no longer needed
Trust based privilege management	Define, assign, and maintain privileges associated with active entities based on established trust criteria consistent with principles of least privilege
Dynamic privileges	Elevate or decrease privileges assigned to a user, process, or service based on transient or contextual factors
Specialisation	Modify the design of, augment, or configure critical cyber resources uniquely for the mission or business function to improve trustworthiness
Protected backup and restore	Back up information and software (including configuration data and virtualised resources) in a way that protects its confidentiality, integrity, and authenticity, and enable restoration in case of disruption or corruption
Surplus capacity	Maintain extra capacity for information storage, processing, or communications
Replication	Duplicate hardware, information, backups, or functionality in multiple locations and keep them synchronised
Predefined segmentation	Define enclaves, segments, or other types of resource sets based on criticality and trustworthiness so that they can be protected separately and, if necessary, isolated
Dynamic segmentation and isolation	Change the configuration of enclaves or protected segments, or isolate resources while minimising operational disruption
Integrity checks	Apply and validate checks of the integrity or quality of information, components, or services
Provenance tracking	Identify and track the provenance of data, software, or hardware elements
Behaviour validation	Validate the behaviour of a system, service, or device against defined or emergent criteria (e.g. requirements, patterns of prior usage)

19 Examples of Potentially Illegal Cyber Security

So where could the methods used to achieve the objectives of these cyber resiliency implementation approaches and the subsequent effect breach the various national and international legal requirements across the world?

Example 1—Deceive Effect

In the first example the asset is protected by the use of deception, where the adversary is operating utilising false information. The false information is provided by the organisation operating the protected asset.

The Deceive Effect is documented below:

20 Deceive

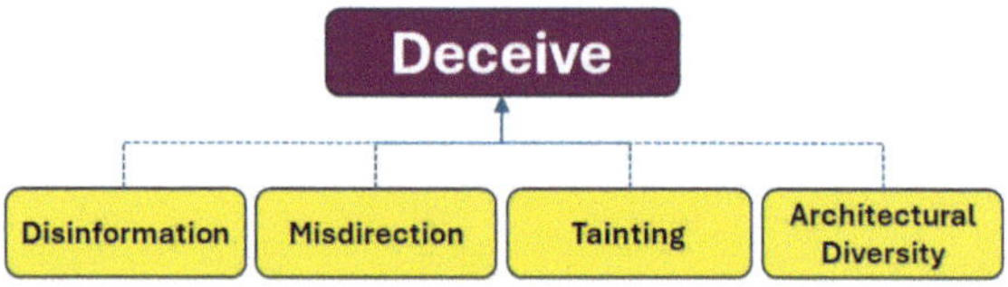

Deceive effect and resiliency implementations

Lead the adversary to believe false information about individuals, systems, missions, organisations, defender capabilities, or TTPs.

This effect is relevant only to adversarial threat events and involves influencing the adversary's actions.

Reduce the likelihood of occurrence, and/or reduce the likelihood of impact.

- The adversary's efforts are wasted as the assumptions on which the adversary bases attacks are false.
- The adversary takes actions based on false information, thus revealing that they have obtained that information.

Example: The defender strategically places false information (disinformation) about the cybersecurity investments that it plans to make. As a result, the adversary's malware development is wasted by being focused on countering non-existent cybersecurity protections.

Example: The defender uses selectively planted false information (disinformation) and honeynets (misdirection) to cause an adversary to focus its malware on virtual sandboxes while simultaneously employing obfuscation to hide the actual resources.

Resiliency approach	Description
Disinformation	Provide deliberately misleading information to adversaries
Misdirection	Maintain deception resources or environments and direct adversary activities there

(continued)

(continued)

Resiliency approach	Description
Tainting	Embed covert capabilities in resources
Architectural diversity	Use multiple sets of technical standards, different technologies, and different architectural patterns

A potential implementation is described below:

Effect:	Deceive
Resiliency Approach:	Disinformation
Candidate Mitigation:	Present Deceptive Information (CM1101) - Present deceptive information about systems, data, processes, and users. Monitor uses or search for presence of that information.
Potential Illegality:	Intentional misreporting of the geographical position that a space asset is in orbit above.

Adversaries may try to identify their target by matching discovered systems with the expected location of the asset they are attempting to attack. Either the system itself, or the information presented about the system, states that the system is not where it is in actuality. The adversary is left in doubt over whether the system they have discovered is the intended target. The hope is that they will not deploy resources against the system where there is doubt over the validity of its identity.

What would be the legal situation if one sovereign nation covertly implemented such a cyber defensive resiliency implementation over another sovereign nation? Should that latter sovereign nation discover the actions of the former, what might their response be? What would the international legal position be, no matter what the national legal positions be?

Although the above example may have legitimacy in a time of conflict between nations, during peacetime it is likely that this could be viewed as an illegal act. Also, could the political fallout be should the two nations be allies?

Parallels may be drawn with the Cold War and the discovery of acts of espionage. However, taking action against systems in orbit is much more complex than expelling Diplomats and detaining those committing such acts.

Example 2—Divert Effect

In the second example the asset is protected by the system redirecting the attack towards another target. This should be invisible to the attacker who remains unaware of the fact that they are attempting to attack the wrong, or a fake, target. As a result the adversary wastes their resources on the redirected attack.

The Divert Effect is documented below:

21 Divert

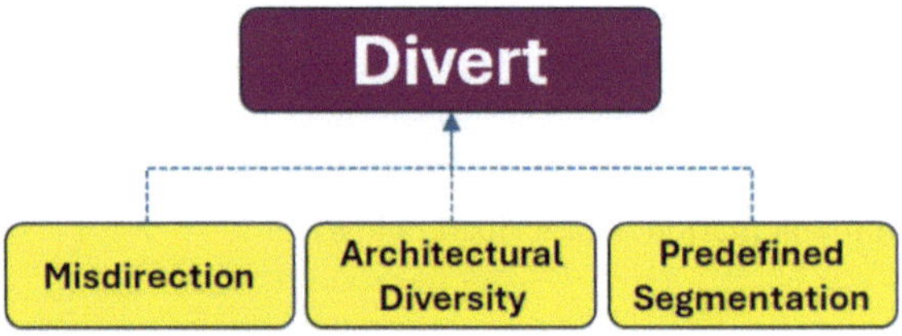

Divert effect and resiliency implementations

Direct the threat event toward defender-chosen resources.

The event affects resources that the defender does not care about or for which the defender can manage consequences.

Reduce the likelihood of occurrence.

- The adversary refocuses activities on defender-chosen resources.
- The adversary directs activities toward targets beyond the defender's purview (e.g., other organisations).
- The adversary does not affect resources that the defender has not selected to be targets.

Example: The defender maintains an Internet-visible enclave with which untrusted external entities can interact and a private enclave accessible only via a VPN for trusted suppliers, partners, or customers (predefined segmentation).

Example: The defender uses non-persistent information and obfuscation to hide critical resources combined with functional relocation of cyber resources and disinformation to lure the adversary toward a sandboxed enclave where adversary actions cannot harm critical resources.

Resiliency approach	Description
Misdirection	Maintain deception resources or environments and direct adversary activities there
Architectural diversity	Use multiple sets of technical standards, different technologies, and different architectural patterns
Predefined segmentation	Define enclaves, segments, or other types of resource sets based on criticality and trustworthiness so that they can be protected separately and, if necessary, isolated

A potential implementation is described below:

Effect:	Divert
Resiliency Approach:	Misdirection
Candidate Mitigation:	Active Decoys (CM1123)—Use one or more factitious systems or other resources to identify malicious sites,

	interact with the adversary, actively probe for malicious code, and observe adversary Tactics, Techniques and Procedures (TTP)
Potential Illegality:	In an intentional misuse of the cyber resiliency implementation approach and the associated candidate mitigation, the defending organisation redirects the adversary to attack a rival's system instead of their own

An adversary tries to attack the control system for a system to disrupt its operation in a denial of service attack. However, the defending organisation has implemented a decoy so that the adversary ends up attacking an alternative system that the defending organisation has redirected them towards. The adversary is unaware that they are attacking a second system, and the original target is unaffected.

If the attack is successful, an unintended and potentially innocent target, at least from the perspective of the attacker, is degraded or disabled. If the parties involved are commercial entities would this constitute industrial sabotage should it be discovered?

Industrial sabotage is defined as deliberate action or inaction that is intended to damage, destroy or disrupt some aspect of the workplace environment, including the property, product, processes or reputation of the organisation (Jermier & Nord, n.d.). Yet, who is the perpetrator and who is the accomplice?

One would assume that the situation is even more complex should the actors involved be nation states. Plausible deniability; intentional entrapment; opportunism. All could play in to such a scenario. What might be the fallout from such a situation?

It is likely safe to conclude that the repercussions are dependent upon who the actors are, should there be any at all.

22 Regulation and Policing

With the two examples above, you may be forgiven for thinking, "So What? Surely this applies whether it is space of terrestrial?".

This is a fair point. However, whereas on land and at lower airspace altitudes there is clear jurisdiction; who has sovereignty in space?

Under the United Nations Outer Space Treaty of 1967,[1] a legal framework was established that looked to ensure that outer space was free for exploitation and use by all states in conformity with international regulations. Within the treaty, States retain jurisdiction and control over objects they have launched into outer space. So who enforces then polices space and takes action in the case of wrongdoing?

[1] The "Outer Space Treaty" Treaty on Principles Governing the Activities of States in the Exploration and Use of Outer Space, including the Moon and Other Celestial Bodies. Adopted by the General Assembly in its resolution 2222 (XXI), opened for signature on 27 January 1967, entered into force on 10 October 1967.

The Outer Space Treaty provides the basis for international space law and includes the following legal principles governing the activities of States in the exploration and use of outer space (United Nations, n.d.):

- The exploration and use of outer space shall be carried out for the benefit and in the interests of all countries and shall be the province of all mankind;
- Outer space shall be free for exploration and use by all States;
- Outer space is not subject to national appropriation by claim of sovereignty, by means of use or occupation, or by any other means;
- States shall not place nuclear weapons or other weapons of mass destruction in orbit or on celestial bodies or station them in outer space in any other manner;
- The Moon and other celestial bodies shall be used exclusively for peaceful purposes;
- Astronauts shall be regarded as the envoys of mankind;
- States shall be responsible for national space activities whether carried out by governmental or non-governmental entities;
- States shall be liable for damage caused by their space objects; and
- States shall avoid harmful contamination of space and celestial bodies.

Of note within these is the statement regarding sovereignty. However, it is also of note that the treaty gives an undertaking that States are responsible for national space activities and are liable for damage caused by their space objects.

Article 7[2] of the Outer Space Treaty covers States' liability to each other; however, this has been elaborated on with the Convention on International Liability for Damage Caused by Space Objects.[3] Whether these international agreements are sufficient to police and govern space with respect to cyber security, both in terms of attacks and/or infringements through the employment of cyber defensive and resiliency implementations, is uncertain. They no doubt could benefit from review being over 50 years old.

The Outer Space Treaty and the Convention on International Liability for Damage Caused by Space Objects are among five treaties, along with five sets of principles, that have been concluded by the Committee on the Peaceful Uses of Outer Space. The committee is administrated by the United Nations' Office for Outer Space Affairs, which 'works to help all countries, especially developing countries, access and leverage the benefits of space to accelerate sustainable development' (United Nations, n.d.).

The five treaties (commonly referred to as the "five United Nations treaties on outer space") concluded by the Committee on the Peaceful Uses of Outer Space are:

[2] "Each State Party to the Treaty that launches or procures the launching of an object into outer space, including the moon and other celestial bodies, and each State Party from whose territory or facility an object is launched, is internationally liable for damage to another State Party to the Treaty or to its natural or juridical persons by such object or its component parts on the Earth, in air or in outer space, including the moon and other celestial bodies."

[3] The Liability Convention was considered and negotiated by the Legal subcommittee from 1963 to 1972. Agreement was reached in the General Assembly in 1971 (resolution 2777 (XXVI)), and the Convention entered into force in September 1972.

- The "Outer Space Treaty"
  - Treaty on Principles Governing the Activities of States in the Exploration and Use of Outer Space, including the Moon and Other Celestial Bodies
  - Adopted by the General Assembly in its resolution 2222 (XXI), opened for signature on 27 January 1967, entered into force on 10 October 1967
- The "Rescue Agreement"
  - Agreement on the Rescue of Astronauts, the Return of Astronauts and the Return of Objects Launched into Outer Space
  - Adopted by the General Assembly in its resolution 2345 (XXII), opened for signature on 22 April 1968, entered into force on 3 December 1968
- The "Liability Convention"
  - Convention on International Liability for Damage Caused by Space Objects
  - Adopted by the General Assembly in its resolution 2777 (XXVI), opened for signature on 29 March 1972, entered into force on 1 September 1972
- The "Registration Convention"
  - Convention on Registration of Objects Launched into Outer Space
  - Adopted by the General Assembly in its resolution 3235 (XXIX), opened for signature on 14 January 1975, entered into force on 15 September 1976
- The "Moon Agreement"
  - Agreement Governing the Activities of States on the Moon and Other Celestial Bodies
  - Adopted by the General Assembly in its resolution 34/68, opened for signature on 18 December 1979, entered into force on 11 July 1984.

The Outer Space Treaty has been reinforced more recently by the establishment of the Artemis Accords. The purpose of the Artemis Accords "is to establish a common vision via a practical set of principles, guidelines, and best practices to enhance the governance of the civil exploration and use of outer space with the intention of advancing the Artemis Program" (NASA, 2020).

The Artimis Accords have come about to support the Artimis campaign to the Moon; and onward to support activities within the vicinity of the Moon and the missions that shall be supported from a permanent presence there. Although a more recently drawn up set of principles, it is based upon collaboration and interoperable working; and although this admiral goal, which brings partner nations together in advancing space exploration, it does not address any further the cyber security threat. Most notably by the fact that neither Russia nor China are signatories.

Yet, infringements need to be addressed. Conversations with persons within the Space Domain have indicated that the likely sanction would ultimately be to remove an assets orbital allocation. How this is enforced remains a question; particularly in the case of a rogue State or an organisation which does not have the resources to recover assets.

The International Telecommunication Union (ITU) under Article 44 of their constitution state that Radio Frequencies (RF) & satellite orbits are limited natural resources. As such they necessitate rational, efficient, economical use through equitable access (Griffin, 2010).

Although they can plan and manage orbital positions and look to being able to guarantee equitable access to orbital resources through predetermined orbital positioning, sanctioning misuse or treaty violations looks more than a hard ask. Instead they talk of a first come, first served approach to provide an efficient orbit management service; rights being "acquired through coordination between administrations concerning actual usage" (Griffin, 2010).

23 Conclusions

Unfortunately, there are no hard conclusions that can be drawn. As said at the start of this chapter, this piece is intended to provoke though and discussion. Space is not sovereign and as such there is no judicial control outside of policing by consent. Is it fair to say that we should ensure that we are a 'good neighbour', and in being so can we then expect other nations to be a good neighbour to us?

We need to be cognisant of our implementations of our cyber resiliency implementations, so as not to potentially infringe or actually break international law, or another nation's laws. It is not the thing I want to be remembered for, having caused a political international incident through the implementation of a cyber defence control.

All of this leaves a number of questions, the answers to be worked out in the future…

How do we enforce international space law and how does it apply to cyber space in a domain with no sovereign control?

Are 40 to 50+ year old treaties sufficient in our modern times?

How can this work if State Users of the Space Domain are not signatories and do not elect to be part of agreements?

Who is going to police it all?

References

Griffin, M. (2010). *Orbit/spectrum allocation procedures*. https://www.itu.int/en/ITU-R/space/workshopBangkok2010/03a-Orbit_Spectrum%20Allocation%20Procedures_MG.pdf

Heukelman, C. (2018). LEO vs. MEO vs. GEO satellites: What's the difference? *Symmetry Blog*. https://www.symmetryelectronics.com/blog/leo-vs-meo-vs-geo-satellites-what-s-the-difference-symmetry-blog/

Jermier, J. M., & Nord, W. (no date). *Pocket: Industrial sabotage*. http://cws.cengage.co.uk/dowling5/students/iebm%20definitions/industrial_sabotage.htm#:~:text=It%20is%20defined%20here%20as,or%20reputation%20of%20the%20organization

MITRE ATT&CK®. (no date). MITRE ATT&CK®. https://attack.mitre.org/

NASA. (2020). *The Artemis accords—NASA*. https://www.nasa.gov/wp-content/uploads/2022/11/Artemis-Accords-signed-13Oct2020.pdf
NIST. (2021). *Developing cyber-resilient systems: A systems security engineering approach*. https://nvlpubs.nist.gov/nistpubs/SpecialPublications/NIST.SP.800-160v2r1.pdf
United Nations. (no date). *United Nations Office for Outer Space Affairs, UNOOSA*. https://www.unoosa.org/

Space Technology and Supply Chain Sustainability: Satellite-Based Due Diligence and the EU Deforestation Regulation

Lisa Heldt

Abstract Today's global supply chains are complex, dynamic and often opaque, with companies struggling to trace back where their material is coming from beyond their direct suppliers and the numerous intermediaries along the chain. Simultaneously, sustainability issues such as deforestation (and the resulting carbon storage and biodiversity loss) but also human rights violations and forced labor tend to be most frequent and devastating deep upstream in the supply chain, effectively hidden from sight. As certification schemes and audits have come under increased scrutiny, companies in forest-risk commodity supply chains (e.g., for palm oil, cocoa, coffee and timber) are now increasingly looking to satellite-based monitoring to gain transparency and continuous oversight over deforestation risk at the raw material source. While this line of application is relatively new, it is gaining momentum as the upcoming EU Deforestation Regulation essentially mandates remote monitoring and new technology providers are entering the market. This chapter explores the current landscape of satellite-based monitoring in supply chain sustainability management, how companies develop the capabilities and collaborations to effectively use satellite data, which challenges and unintended side effects emerge as well as where the opportunities and pitfalls lie when looking ahead, in terms of lasting sustainability impact on the ground, collaboration across the supply chain as well as interoperability and information security questions.

Keywords Supply chain management · Sustainability · Satellite monitoring · Due diligence · EUDR · Traceability · Ethical concerns

L. Heldt (✉)
International Institute for Industrial Environmental Economics (IIIEE), Lund University, Tegnersplatsen 4, P. O. Box 196, 22100 Lund, Sweden
e-mail: lisa.heldt@iiiee.lu.se

H. Jahankhani et al. (eds.), *Space Law Principles and Sustainable Measures*, Space Law and Policy, https://doi.org/10.1007/978-3-031-64045-2_6

1 Introduction

Space technology and particularly satellites have revolutionized our understanding of Earth and its ecosystems: From mapping land use changes to monitoring climate patterns, satellites have become indispensable in environmental monitoring, resource management and disaster response on a global scale. In recent years, the application of satellites for Earth observation has extended beyond scientific and national domains and commercial applications in the context of supply chain management and sustainability are on the rise.

Today's global supply chains are complex, dynamic and often opaque, with companies struggling to trace back where their material is coming from beyond their direct suppliers and the numerous intermediaries along the chain. Simultaneously, sustainability issues such as deforestation (and the resulting carbon storage and biodiversity loss) but also human rights violations and forced labor tend to be most frequent and devastating deep upstream in the supply chain, effectively hidden from sight. Research suggests that "the further upstream an organization is the more impact it is likely to have on sustainability" (Mena et al., 2013, p. 72). As traditional approaches such as certification schemes and audits have come under increased scrutiny (Marin-Burgos et al., 2015; Schepers, 2010), companies with complex supply chains are now increasingly looking for other solutions to bridge the visibility gap and get more direct oversight. Companies with forest-risk commodity supply chains are increasingly looking to satellite-based monitoring to gain transparency and continuous oversight over deforestation risk at the raw material source (Heldt & Beske-Janssen, 2023).

Coming from a sustainability and management background with no technical knowledge in remote sensing, satellite monitoring of deforestation seemed to me—quite literally—a solution from space. When I started researching this phenomenon in 2019, using space technology for earth observation, specifically monitoring deforestation in supply chains, was a relatively new application and used to be limited to some leading companies, running pilots and trying to figure out what they needed from the solutions and how to talk with technology providers about this. Now, however, it is gaining momentum since the EU Deforestation Regulation (EUDR) entered into force in mid-2023 (European Commission, 2023c) and essentially mandates space-based monitoring of deforestation: The EUDR requires companies to conduct due diligence and demonstrate evidence that forest-risk commodities (i.e. palm oil, coffee, cocoa, cattle, rubber, timber and soy) placed on the EU market are not linked to deforestation. So far, so good. Now, the EUDR requires companies to provide full traceability of material back to the raw material source, as well as geolocation data of said farms, forest and/or plantations, plus regular monitoring of these locations—which, considering the costs, time and efforts involved in in-person audits, puts the spotlight on space-based monitoring solutions. Whereas some leading companies have explored and matured their use of satellite-based deforestation monitoring, many companies affected by EUDR are now scrambling to understand and navigate this space as technological possibilities and providers keep increasing. The insights discussed in this chapter span this research period, from previously published results (Heldt, 2020; Heldt & Beske-Janssen, 2023) to ongoing investigations of EUDR implications for satellite technology use.

This chapter explores the current landscape of satellite-based monitoring in supply chain sustainability management, how companies develop the capabilities and collaborations to effectively use satellite data, which challenges and unintended side effects emerge. We will cast a look ahead to understand where the opportunities and pitfalls lie moving forward, in terms of lasting sustainability impact on the ground, collaboration across the supply chain but also questions relating to interoperability, information security and ethical concerns of using space technology to drive sustainability in supply chains.

This book is a timely and much needed cross-disciplinary platform to discuss the nexus between space technology and sustainability with inputs both from space law and information security experts as well as researchers and managers with expertise on the application side. I hope the book kickstarts the conversation and draws more attention to the wider implications and potentials of space technology, both for driving supply chain sustainability and wider societal transformation but also regarding security and ethical concerns that need to be considered. In this context, my chapter provides a practice-based perspective on how companies have started to adopt satellite monitoring in their efforts to achieve supply chain sustainability and reduce impacts such as deforestation, climate change and biodiversity loss.

2 Background

2.1 Supply Chain Sustainability

This section introduces supply chain management and supply chain sustainability, exploring how today's global supply chains are structured and managed, which issues they face and why sustainability is a growing concern.

Supply chains are defined as "the organizations and flows of material, capital and information between organizations from raw material extraction to end-users" (Halldórsson et al., 2007, p. 286). Today's globalized supply chains resemble networks rather than chains, with multiple, so-called 'tiers' of suppliers extending from the focal company upstream, i.e. towards the raw material source (see Fig. 1). Traditionally, supply chain management, both in research and practice, has mostly focused on focal companies' relations with direct suppliers (Choi & Wu, 2009), perhaps counterintuitively since the term supply *chain* management implies a more comprehensive, networked perspective (Dubois, 2009; Tachizawa & Wong, 2014).

Environmental and human rights scandals over the last decades have given rise to sustainable supply chain management (SSCM), also called sustainable sourcing (note: they have different nuances, but for the sake of this article, you can treat the terms interchangeably). SSCM is concerned with how to manage supply chains

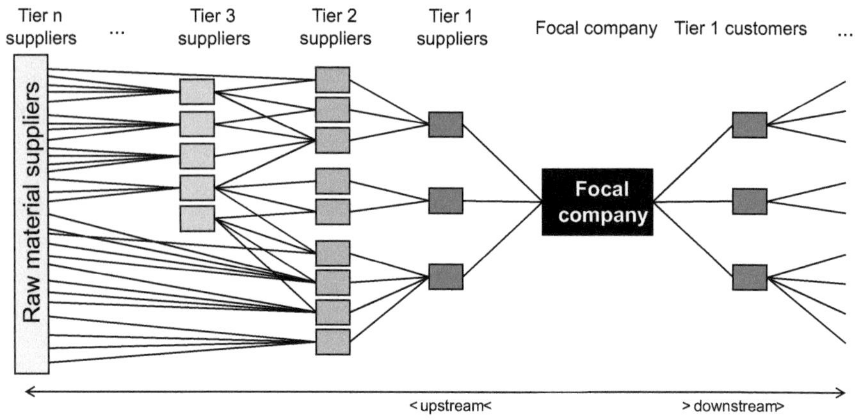

Fig. 1 Simplified supply chain structure and key terms (based on Heldt, 2020)

while taking "all three dimensions of sustainable development, i.e., economic, environmental and social, into account which are derived from customer and stakeholder requirements" (Seuring & Müller, 2008, p. 1700). This entails setting minimum sustainability standards for suppliers, e.g. supplier code of conducts, verifying sustainability performance through audits or working with suppliers through capacity-building and supplier development and possibly collaborating in multistakeholder initiatives to address more systemic issues (Akhavan & Beckmann, 2017).

Companies' sustainability efforts have mostly concerned their direct first-tier suppliers because they typically do not have direct contractual relations, communication channels or leverage with indirect suppliers, i.e. issues are complicated to address. From a sustainability standpoint, however, issues tend to be most severe and likely to occur far upstream, beyond focal firms' visibility. Think of child labor in cobalt mining, deforestation in agricultural supply chains or water shortages due to cotton production for textile firms. These common issues occur several tiers deep in the supply chain where they are effectively hidden from sight. Simultaneously, this is where ecosystems and communities are more vulnerable, e.g. indigenous peoples without proper representation or tropical rainforests that represent biodiversity hotspots. Further, issues are more likely to occur and persist where institutions and legal frameworks are immature or rendered weak by corruption and political instability. Thus it can be said that "the further upstream an organization is the more impact it is likely to have on sustainability" (Mena et al., 2013, p. 72). Thus, companies are increasingly working to increase visibility into their supply chain, i.e. identifying and mapping their upstream suppliers, as well as traceability, i.e. the ability to track specific products back to their upstream origin and establish provenance (Sodhi & Tang, 2019).

Essentially, sustainability issues tend to be more severe, frequent and persistent when they occur far upstream, and in recent years, companies have come under increasing pressure to adopt a multi-tier approach to supply chain sustainability—or face business repercussions (Gong et al., 2023; Jamalnia et al., 2023). Whereas upstream sustainability issues were (and continue to be) difficult to address, the tide—or incentive structure—is turning. Global supply chains are still complex, dynamic and opaque but companies face growing business risks from not knowing or acting on what is happening in their chain. Four risk categories can be differentiated:

To sum up, today's global supply chains are complex, dynamic and opaque. It is difficult for focal companies to know and manage what is going on beyond direct suppliers, but not impossible. Sustainability issues far upstream tend to be more severe, frequent and persistent than issues at direct suppliers. Recently, business risks from inaction are growing and companies face an incentive to create visibility and leverage deep upstream—yet, it is often unclear how to achieve this.

2.2 *Satellite Technology for Earth Observation*

Satellite-based Earth observation is no new technology per se yet has experienced tremendous momentum over the last years and transformed our ability to monitor the planet's ecosystems, climate, infrastructure and land use in an unprecedented granularity. These advancements have been driven by a combination of technological innovations, cost reductions in satellite manufacturing and launch, and improvements in data processing and analysis capabilities—but also tapping into new application markets.

Essentially, Earth observation or satellite-based remote sensing are about taking pictures of the planet from space and then analyzing this satellite imagery for various purposes. Remote sensing is "*the use of aerial or satellite imagery to study features on Earth's surface*" (Werner et al., 2019, p. 994) and, combined with geographical information systems (GIS) "*by which geographical data are managed, analyzed and manipulated, usually to produce visualizations (2-D or 3-D), including maps*" (Werner et al., 2019, p. 994) it allows for relatively direct, continuous observations of environmental change. Whereas governments, specifically defense, and research are the original and primary users of satellite monitoring, this has gradually expanded to commercial applications, e.g. spatial planning or vegetation management for infrastructure or railroad companies and—most recently—for monitoring deforestation and carbon stocks (Csillik et al., 2019; Hansen et al., 2013).

2.2.1 Data Collection

Historically, the high cost of satellite manufacturing, launch, and operation has been a barrier to widespread adoption of satellite-based monitoring systems. With a shift towards smaller satellites and using standardized off-the-shelf components which drives down cost, there are more and more satellites being launched, including commercial satellite constellations. Similarly, spectral resolution has improved and data is now available from high-resolution optical sensors plus radar detection and ranging (Radar) sensors, especially synthetic aperture radar (SAR), and light detection and ranging (Lidar) sensors. Using different wavelengths along the electromagnetic spectrum, these different sensors enable a granular mapping of the planet and surface structure. Overcoming limitations of optical satellites when it comes to piercing cloud cover, forest canopy or darkness is important for getting accurate pictures for the most vulnerable, climate-relevant and biodiversity-rich rainforests that often face cloud coverage. Lidar's capabilities in capturing vegetation structure below forest canopies enables estimating aboveground biomass and carbon stock accounting for climate targets, demand for which can only be expected to grow. Besides spectral resolution, there has been much progress in temporal resolution, i.e. "*the time period between repeat passes over an object being remotely sensed*" (Turner et al., 2003, p. 307), which for satellites like Planet's constellation is down to daily or even sub-daily revisit times (Planet, 2020) thereby tackling cloud cover or darkness obstructions from a different (optical) angle. Similarly, spatial resolution of the images produced has gone down from the 30m resolution of NASA's Landsat data to a 50cm resolution with Planet's fleet (Planet, 2020). In the defense and intelligence space, Albedo Space is even planning for 10cm resolution which sparked concerns over privacy and ramifications for human rights (Broad, 2024). While happening outside the domain of forest monitoring, it foreshadows similar discussions that may emerge around ethical concerns and misuse of technological advances. But overall, as new, including commercial, satellite constellations are launched and drive down prices for high resolution imagery, this makes satellite monitoring attractive and affordable for different applications where detailed imagery is needed at regular periods.

2.2.2 Data Analytics

Besides improvements on the data side, the rise in cloud computing and sophisticated algorithms now enable faster and more accurate data processing (Sisodiya et al., 2020) to turn raw data into actionable insights like maps or automatic alerts. This is particularly important because it enables handling data at scale and dealing with the increasingly large datasets available through improvements discussed above in relation to data collection. Previously, data analytics and change detection, e.g. in forest cover, used to be conducted manually but relying on algorithms based on machine learning or artificial intelligence enables much more efficient processing (Sisodiya et al., 2020). Simultaneously, better analytics does not only help to handle more data. It is also relevant because it—to an extent—can compensate for better data, considering that much of the higher resolution imagery is proprietary. Improvements in machine learning or artificial intelligence can enable algorithms to get more insight out of coarser public data. Similarly, more expensive datasets, e.g. LiDAR data, can be used to train algorithms and specifically validate and improve, e.g. their detection of forest degradation or mapping of carbon stocks, e.g. Planet or Chloris Geospatial, letting them transfer this to the analysis of optical satellite data moving forward. With the proliferation into a range of client industries and applications, the technology provider market is growing and specializing and with it the tailored algorithms and platforms that many build their competitive advantage around.

In summary, there is a flurry of more satellite constellations, better resolutions, more efficient and capable data processing capacities, more sophisticated algorithms, increasing affordability and a growing range of industries and companies discovering the possibilities of satellite monitoring. Things are moving fast and the emerging opportunities are fascinating. The next sections explore this in more detail for the application for monitoring sustainability issues in supply chains and also dives into unintended side-effects or ethical concerns that may arise.

3 Satellite Tech for Supply Chain Deforestation Monitoring

3.1 Drivers

Adoption of satellite technology for corporate deforestation monitoring is comparatively recent and the first wave of adoption is pushed by four central drivers (Heldt, 2020): risks from inaction, frustration with previous approaches, technological accessibility and growing commercial opportunities from using satellite data.

First, the risk of inaction or insufficient action is growing. We already discussed how sustainability issues can spiral into reputational, legal, operational or financial risks for companies (see Table 1). The stakes are getting higher as companies face repercussions for inaction and being associated with deforestation by stakeholders.

Second, frustration with weaknesses of existing approaches is growing. So far, companies have traditionally been relying on certification schemes, such as the Forest Stewardship Council (FSC) or the Roundtable for Sustainable Palm Oil (RSPO) certification, to source verified deforestation-free products, essentially outsourcing the standard setting, certification and auditing to third parties. Yet, audits have known limitations, especially for monitoring large, remote forested areas: in-person audits are time- and resource-intensive for both sides, they represent snapshots of one point in time rather than reliable or continuous representations and they can be vulnerable to corruption and partiality. Scandals around certification schemes have shaken not just consumers but also companies' trust in the labels (Lopatin et al., 2016), making them look for alternatives or complementary approaches for assurance.

Third, the accessibility of satellite technology is increasing. For a long time, satellite technologies were quite literally 'solutions from space' and not necessarily the first thing that comes to mind for a manager seeking to reduce sustainability risk exposure. With advances in technology and its commercialization as discussed in the previous section, satellite imagery and derived platforms and projects are becoming increasingly available, affordable and thus accessible to corporate customers.

Fourth, operational improvements and business opportunities. Satellite-based monitoring enables companies to take a more efficient and effective approach with

Table 1 Business risks emerging from supply chain sustainability issues (Heldt, 2020, p. 7)

Risk type	Explanation
Legal and regulatory risk	New mandatory Human Rights and environmental due diligence (mHREDD) regulations are increasingly holding companies legally accountable for poor due diligence concerning illegal logging (Partzsch & Vlaskamp, 2016) or conflict minerals (Hofmann et al., 2018) in their upstream supply chain
Reputational risk	Customers, NGOs and the public expect buyers to assume responsibility for mitigating issues like forced labor or deforestation associated with their products (Lozano, 2015; Roberts, 2003). That is, even when focal companies have outsourced production and processing steps to suppliers, the perceived responsibility for environmentally and socially sound production standards remains with them (Foerstl et al., 2010);
Operational risk	Environmental issues or extreme weather events can disrupt suppliers' operations and thereby cause supply bottlenecks, quality issues or price spikes (Hofmann et al., 2014). Under the influence of climate change, both the intensity and frequency of extreme natural events like floods or droughts is expected to increase, adding further pressure to operational planning, sound risk management and adaptation measures (Ghadge et al., 2020)
Financial risk	Investors and banks recognize that supply chain sustainability issues can threaten business performance and thus seek to remove these risks from their portfolios, which can limit access to capital (Tropical Forest Alliance, 2017). Likewise, clients of business-to-business firms increasingly seek to reduce risk in their own supply chains and therefore demand sound sustainability standards as a contractual condition (Beske et al., 2014)

their auditing and supplier development processes: Screening suppliers based on satellite data helps identifying risk areas and where to target additional in-person audits or targeted measures to address problems as a way to use resources more effectively. Further, there is a commercial incentive to overcome silo-thinking and use insights from satellite data not just for no-deforestation efforts but instead connect it with the sourcing department's insights and efforts. Adopting an integrated data-driven approach can thus inform companies' sourcing on how to develop farmers to achieve better yields and/or become more climate resilient, and thus secure long-term access to quality supply of e.g. cocoa or coffee which are under severe threat from climate change and likely to incur supply disruptions and price spikes (Ghadge et al., 2020).

3.1.1 The EU Deforestation Regulation (EUDR)

While building on or reinforcing these underlying drivers, the EU Deforestation Regulation's entry into force in 2023 has eclipsed other factors. The EUDR can be viewed as the decisive impulse for the second, broader wave of companies now scrambling to adopt satellite monitoring of deforestation. Especially since corresponding legislations in the UK, USA and China appear to be underway.

Officially called the*'Regulation (EU) 2023/1115 of the European Parliament and of the Council of 31 May 2023 on the making available on the Union market and the export from the Union of certain commodities and products associated with deforestation and forest degradation and repealing regulation (eu) no 995/2010'* (European Commission, 2023c), it is commonly known as the EU Deforestation Regulation or EUDR. It requires any company that puts forest-risk commodities on the EU market to conduct due diligence and provide evidence that production has not been connected to deforestation or forest degradation (after the 31 December 2020 cut-off date, Art. 3a). The commodities in scope are cattle, cocoa, coffee, palm oil, rubber, soy and wood, plus derived products like leather, books or chocolate. Besides the requirement of no-deforestation, the commodities also need to be produced legally, i.e. respecting local laws concerning labor and human rights, environmental protection or anti-corruption and tax laws (Art. 3b). Compliance is checked by so-called Competent Authorities appointed by EU Member States and non-compliant companies risk fines of up to 4% of their previous year's EU revenue, seized goods, plus exclusion from public tender processes—and the resulting reputational damage and supply delays.

To demonstrate EUDR compliance, companies need to implement a due diligence approach consisting of three steps: Information requirements, risk assessment and risk mitigation.

(1) **Information requirements (Art. 9).** Companies need to be able to trace these commodities back to the each specific plot of land where they were produced and collect geolocation data of the plot of land (polygons for plots > 4 hectares) along with information concerning quantity, date of production, suppliers etc. and provide this information in a due diligence statement.
(2) **Risk assessment (Art. 10).** Companies need to verify and analyze the information collected in the previous step and assess the risk that the respective product is non-compliant with EUDR. The law specifies a number of criteria to consider, such as the prevalence of deforestation generally or concerns around corruption, data falsification or lack of law enforcement in a country. The risk assessment needs to be repeated on an annual basis (at least).
(3) **Risk mitigation (Art. 11).** Companies need to put in place adequate risk mitigation procedures on how to react in case they detect a risk in their Step 2 risk assessment and how to mitigate and manage this risk to achieve a level of 'no or negligible risk'. This could include collecting additional data or conducting independent audits.

The EUDR entered into force on 30 June 2023 with implementation due 18 months later (30 December 2024) for large and medium-sized companies and 24 months for small- and micro-sized companies. The EUDR is probably the most ambitious, comprehensive and stringent regulation on deforestation globally to date and is sending a strong signal—one of the key reasons why companies are looking to data-driven approaches and satellite monitoring for their risk assessment. While the EUDR's requirements are in line with what most more advanced companies have been doing, it entails large potential to pressure laggards into action and thus raise the floor. What still makes it challenging to conduct due diligence, even for leading companies, is that the requirements are in many regards absolute and grant limited leniency.

- Supply chain complexity or opacity (and, essentially, this is the point of the regulation) do not excuse companies from demonstrating full traceability. The requirements apply uniformly"*regardless of the length or the complexity of [the] supply chains (...) For instance, a large, bulk shipment of soy that has been sourced in several hundred plots of land from several countries would need to be associated with a due diligence statement that includes all relevant countries of production and geolocation information for every single plot of land from all of these countries that has contributed to the shipment*" (European Commission, 2023b, p. 11).
- Traceability and 'no or negligible risk' status needs to be demonstrated for the entire batch and e.g. 98% traceability is not enough. The marginal effort for achieving these last percentage points is substantially higher. Similarly, there is no lower threshold for companies that import small amounts or have forest-risk commodities as a minor ingredient in a product: The EUDR contains "*no threshold volume or value of a relevant commodity or relevant product, including within processed products, below which the Regulation would not apply*" (European Commission, 2023b, p. 13).
- The EUDR requires non-mixing of EUDR-compliant and potentially non-compliant materials (Art. 10 (2)). This means that materials cannot be mixed at any point, consequently, implying separate infrastructures for mills, refineries, siloes or transport.

Previously, companies have been relying on environmental certifications like FSC or RSPO to ensure produce is not linked to deforestation, essentially using the certification's standards and third-party assurance processes as proxies for conducting their own investigations. Under the EUDR, certification is not accepted as evidence of no-deforestation. While initially unclear, the EU clarified that certification can support risk assessment, but ultimate responsibility for compliance continues to lie with the company. At the time of writing (Spring 2024), further such clarifications as well as the EU's release of its guidance and expansion of the Frequently Asked Questions are pending and, in light of the December 2024 due date, urgently needed.

To comply with these tight requirements, companies are looking for reliable ways to conduct the information collection and risk assessment and protect themselves against non-compliance. So far, companies' no-deforestation efforts have been technology- or approach-agnostic. But the EUDR's no-leniency approach, steep fines and regular risk assessments require comprehensive, continuous and data-driven solutions, essentially spelling satellite-based monitoring. The risk assessment step is where satellite monitoring emerges as an intriguing solution: in contrast to expensive in-person audits of all plots, satellite imagery appears to promise continuous, almost real-time, reliable, unbiased assurance from the comfort of your computer. Of course, this is oversimplified but to corporate sustainability managers who have never before come in contact with remote sensing or satellite imagery beyond, perhaps, the satellite view on Google Maps, it is both an attractive but also intimidating technology that is hard to grasp. The next section takes a closer look at how to integrate the satellite imagery with the surrounding SSCM approach and how to embed the EUDR's due diligence system effectively.

3.2 Application of Satellite Technology for Deforestation Monitoring in SSCM

The adoption of satellite technology in SSCM (before and after EUDR entry into force) is essentially about gaining objective and direct information and enabling more data-driven decision-making for the notoriously opaque context of supply chain sustainability. We discussed the different reasons why companies are seeking more reliable, direct and continuous insights in the previous section, so let us now look at how they are doing this.

The basic approach is to layer current satellite imagery against datasets such as the baseline level of forest cover or land use to detect changes or deforestation events using specialized algorithms. To make this relevant to supply chains, it is important for buyers to be able to understand if suspected deforestation is connected to their own suppliers. For this, additional data is needed. Ideally, concession boundaries and polygons of suppliers' plantation or farm boundaries are available from previous efforts to establish visibility into the supply chain. This data can be contrasted with detected forest cover change to understand which specific supplier may be associated. If such granular datasets are not available or upstream suppliers are unknown, coarser estimates can be used which I will cover more below. Further dataset layers can be added to provide a differentiated understanding of situation and context: Such datasets can include protected areas like national parks, areas of high conservation value or high carbon stock, peatlands, or datasets on land rights and cadastral information. These can then be contrasted with maps of suppliers' plots and deforestation events to ensure there is no violation of land rights or deforestation outside concessions. Which datasets are used depends on commodity and regional dynamics, their availability and possible integration into the different tools.

Companies usually work with a technology provider to gain these insights. The technology providers typically exist on a spectrum from focusing on data collection and provision with own satellite constellations with very-high resolution imagery to providing processing platforms or tailored analytics solutions that are highly customized to a particular sector or impact, e.g. Satelligence's solutions for agrifood supply chains or Airbus and Earthworm Foundation's Starling service. Overall, the spatial, spectral and temporal resolution of imagery will vary, the ability of algorithms to detect not just clear-cut deforestation but also forest degradation or small-scale deforestation events within intact forests, the surrounding support on what is needed for EUDR compliance or how to engage with suppliers and stakeholders. Which provider is best suited to a particular company depends on a variety of factors, such as their internal technical skills, their SSCM expertise and their ambitions. The next section will cover this more closely.

4 Integration of Satellite Technology Into Sustainable Supply Chain Management

This section introduces how satellite monitoring has been and needs to be embedded in companies' broader SSCM practices in order to manage deforestation risk on the ground and the resulting legal, reputational, operational and financial risks to the business effectively. The premise of using satellite imagery to spot deforestation sounds straightforward and simple enough. Yet, getting from raw data to actionable insights and positive on-the-ground impact can be a challenging journey, especially for companies taking their first steps in this space. Based on insights gained from early adopters of satellite technology for SSCM, this section attempts to demystify what companies need and what technology providers and space professionals need to understand about their clients' needs, challenges and broader SSCM context when embarking on the satellite tech journey.

4.1 Integration With SSCM Approach

As introduced before, today's global supply chains are incredibly complex. For companies to gain meaningful, actionable insights from satellite imagery, they need to be able to identify where their suppliers are located in order to map those locations against baseline forest data and deforestation alerts. Establishing full traceability, all the way to the beginning of the raw material source, through various traders, refineries, exporters and mills is a tremendous challenge. Large companies (think of the parent company behind your favorite mainstream chocolate brand) typically source their cocoa (via intermediaries and traders) from hundredthousands of small-holder farmers, each cultivating an area of only a few hectares. And cocoa is of course only one part. Typically, the same companies also source palm oil, coffee and other forest-risk commodities. Getting full traceability does not happen overnight.

In previous research (Heldt, 2020; Heldt & Beske-Janssen, 2023), we explored how companies that were pioneering satellite-based deforestation monitoring before EUDR integrated the insights gained into their overall SSCM approach. We found that, to effectively manage sustainability and business (legal, reputational, operational, financial) risks, companies needed their SSCM to fulfill four functions: traceability, monitoring, follow-up and accountability (see Fig. 2). Whereas we already covered the 'pure' monitoring aspect earlier in this chapter, the surrounding three functions are critical for satellite imagery to provide actionable, valuable insights in

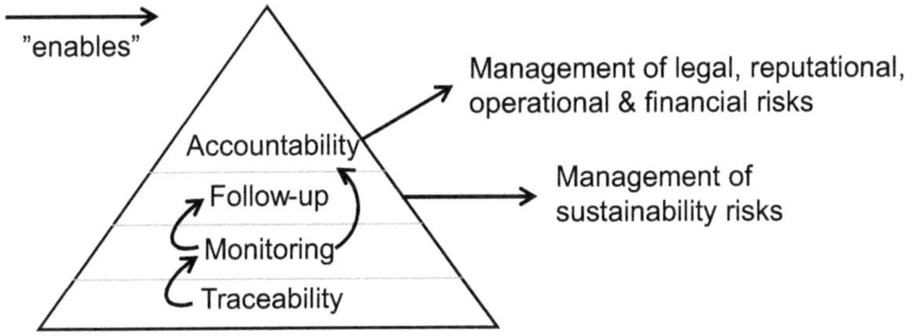

Fig. 2 Functional integration of satellite monitoring technology into sustainable supply chain management (Heldt & Beske-Janssen, 2023)

companies. And as companies, especially those thrust into this discussion through the EUDR, are often unfamiliar with satellite technology and what exactly they need when looking for the right technology provider, I hope to contribute to the mutual understanding with this chapter and clarify to technology providers what the situation looks like for their clients.

Traceability is the foundation for being able to monitor your raw material suppliers. This is reflected in the EUDR's first step in the due diligence process which concerns information requirements and (geolocation) data collection. This may sound trivial and straightforward, but remember that in reality, supply chains are complex, dynamic and opaque and focal firms often do not know their chain beyond their direct suppliers. The structure of agrifood commodity supply chains entails that products flow through, e.g. mills or domestic exporters, where material from different suppliers gets mixed and traceability to origin gets lost. Further, agrifood supply chains, especially cocoa and coffee but also to an extent palm oil, entail varying amounts of smallholder farmers besides large plantations. Smallholders with a few hectares each means more suppliers, more data points and more efforts to establish traceability. Yet, smallholder inclusion is an important element of most companies' sustainability strategy to improve livelihoods in sourcing regions and avoid marginalizing an already vulnerable group that often pushed into deforestation out of poverty or poor agricultural practices that require more land (Smit et al., 2015). Establishing traceability to farm, plantation or forest-level may actually be the hardest part in getting to EUDR compliance for companies that are starting from scratch and have not engaged with mapping their supply chain and trying to increase their upstream visibility at all before.

The companies we have followed as early adopters (Heldt, 2020; Heldt & Beske-Janssen, 2023) have built traceability over years. This is one example for why the EUDR can be considered a regulation that is raising the floor, not the bar, on sustainability efforts, pushing for broader adoption of anti-deforestation efforts in industry. On this journey, besides working with suppliers to get them to share data on their upstream suppliers (which suppliers can be wary of, considering the data confidential or fearing to be leapfrogged), companies can adopt risk-based approaches and work with proxies to narrow down where to focus attention for building visibility and collecting geolocations. For instance, depending on the commodity and region, raw material can only be transported a certain distance before, e.g. oil palms' fresh fruit bunches need to be processed or before transport becomes uneconomical. Thus, you can identify the maximum transportation radius around a mill to estimate its sourcing region. Similarly, datasets like concession boundaries or national parks can be added to identify suppliers and mills close to protected areas where deforestation would be more serious. While this does not provide actionable alerts, it does help companies identify where to prioritize efforts to identify raw material producers. Other examples include letting transportation providers opt-in to have their trucks location GPS-tracked or tracking anonymized mobile phone data to link plantations to mills via drivers. Building traceability in supply chains is no quick exercise and may require some creativity, but it pays off and regardless of whether satellite monitoring is implemented for EUDR compliance or as support for sustainability efforts more generally, traceability is the precondition for being able to make deforestation alerts actionable by tracing them to specific suppliers.

Follow-up on suspected deforestation issues is critical to establish whether the deforestation alert is correct and to decide on further actions. Satellite imagery and/or algorithms for deforestation detection are not perfect and may flag 'false positive' where e.g. forest cover loss is due to natural fire (assuming the area is not converted into agricultural land use after) or where poor imagery creates a false baseline, e.g. replanting on a smallholder farm being flagged as deforestation because the baseline map identified the farm as forest. Even the EU Forest Observatory (2023) map of 2020 forest cover has been criticized for inaccurately mapping perennial crops (e.g. cocoa or oil palm) and smallholder farms (Satelligence, 2024), raising question marks around which role it will (or can) play in EUDR compliance (see Sect. 5.2.3 for details). Coming back to the topic, follow-up on deforestation alerts is critical to verify what is going on on the ground. Contacting suppliers for further information (again, having full traceability and good lines of communication helps) is a first step and where needed, this is combined with field visits. Again, focal companies that are far removed from raw material suppliers may rely on their traders or technology providers who have local offices or liaisons for ground truthing efforts.

When it comes to the EUDR, once companies detect deforestation, they cannot place this material on the EU market anymore. While post-detection follow-up and remediation is thus (interestingly) out of scope of the EUDR, preemptive follow-up or field visits can be useful or necessary to support the due diligence process for assessing deforestation risk (and legality under local laws). So while the EUDR is essentially about providing 'no deforestation' evidence, this is only one part of the puzzle of overall supply chain sustainability. More broadly speaking, companies need to have procedures in place for how to deal with suspected or detected deforestation issues—procedures that go beyond simply redirecting risky sources to non-EU markets or terminating supply relations without attempting to investigate root causes or solving (rather than avoiding) the problem. Deforestation, especially when linked to smallholders, can have underlying causes like poverty, inefficient agricultural practices that lead to poor yields and necessitate expansion, unclear land rights or even be considered legal deforestation in the country of origin, creating a complex situation that may require more farsighted measures than terminating supply relations. Further, deforestation can co-occur with child labor, unsafe working conditions or forced labor, and a forced stop on deforestation may intensify these connected issues by straining the plantation or farm's financial situation further. So follow-up on deforestation alerts needs to ideally be linked with broader assessments of ground conditions and root causes, e.g. using solutions like Ulula's worker voice tool to collect complementary data on human and labor rights issues, and derive measures that address rather than aggravate root causes. For this reasons, many leading companies increasingly have smallholder inclusion and fair livelihoods as objectives alongside no-deforestation or other environmental targets. Lastly, although terminating supply relations should be seen as a 'last resort' when the goal is to maintain leverage and create lasting change on the ground, there are of course cases where terminating the supply relation is the only sensible option, e.g. in cases of locally legal deforestation where suppliers do not see the point in refraining from clearing the land.

Accountability to stakeholders is important for companies in order to reduce their risk exposure from stakeholders fearing sustainability issues. Often, sustainability issues on the ground only create business repercussions when they are taken up and raised by stakeholders, such as NGOs but also investors or regulators. Environmental certification used to be a classic option for companies but has itself come under intense scrutiny and criticism. Simultaneously, stakeholder engagement is a central part of good corporate sustainability strategies and the EUDR stipulates that as part of their due diligence system, companies should report annually on their due diligence approach to stakeholders. This means companies that focus on marketing and communication (or avoiding communication altogether) may have gotten away with poor on-the-ground conditions, at least pre-EUDR. On the other hand, companies with a good track record and no deforestation need to be able to convincingly convey this to stakeholders, e.g. in their EUDR due diligence statements, annual reports and shareholder meetings. Solving issues silently—unfortunately—does not reduce companies' risk level. Similarly, under the EUDR, if the due diligence and reporting is not clear and convincing enough, products may be held up at customs even if there is no deforestation on the ground, depending on how the Competent Authorities interpret the law. Again, accountability and convincing stakeholders is critical for minimizing business risks.

4.2 Collaboration Across Supply Chains

Now, the functions discussed above are at times challenging to obtain/provide internally. Companies' skills and expertise may vary widely, depending on how long they have been seriously pursuing supply chain sustainability, how mature their SSCM system is (e.g. how much upstream visibility do they have?) or which industry they are in. Even just having a team member in the sustainability department who happens

Fig. 3 Governance mechanisms in technology-enabled multi-tier SSCM (Heldt & Beske-Janssen, 2023)

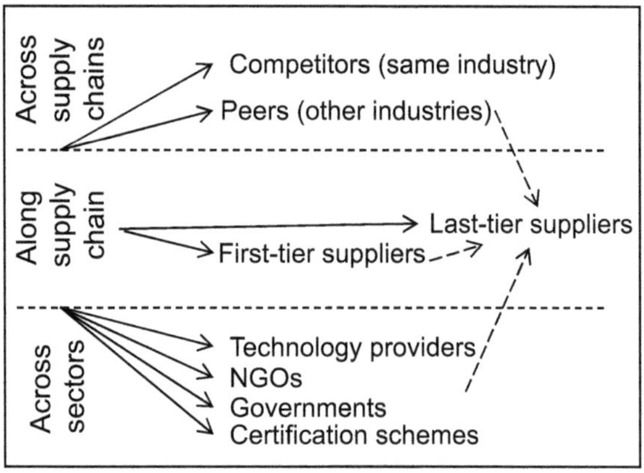

to have a remote sensing background makes a difference. To compensate for internal weaknesses, companies collaborate selectively with partners to compensate and three main governance mechanisms for these multi-tier collaborations can be distinguished (see Fig. 3).

Linking back to the previous section, you can see how these different governance mechanisms play out in practice. For instance, focal companies that are far removed from their sourcing regions (operationally, geographically and/or culturally) may **work along their supply chain** and seek support from intermediary suppliers like large traders that have more technical expertise inhouse and more direct upstream relations when it comes to collecting geolocation data from upstream suppliers or helping smallholder farmers with polygon mapping. Companies may also decide to **work across supply chains**, either with competitors operating in the same industry or with peers in other industries (e.g. food&beverage companies and personal care companies that both source palm oil for their products) that source from the same region in order to pool resources and have more leverage when addressing deforestation issues. Similarly, when it comes to following up on a deforestation alert and companies lack understanding of or access to the local context, companies may **work across sectors** and collaborate with local NGOs in order to understand root causes that drive deforestation and how to prevent it. Companies may also combine approaches and engage with competitors sourcing from the same regions as well as local governments and other stakeholders to address more systemic issues in so-called jurisdictional or landscape approaches. In these multi-stakeholder settings, satellite data has been pointed out as useful for establishing common ground and agreement between actors as a data source that is (considered/accepted as) 'objective' or 'neutral', again supporting accountability to stakeholders (Fig. 2). These landscape approaches are also critical to prevent so-called 'leakage' (Godar et al., 2016, p. 10) of deforestation that occurs when deforestation is narrowly prevented for a particular commodity and place but the deforestation pressure is essentially just pushed onto adjacent areas or other commodities within the broader production landscape. Similarly, there can be a leakage of deforestation-tainted product onto the market when a company terminates relations with a supplier for deforesting but then other less scrupulous buyers purchase the same material. Systemic approaches are needed for effective, lasting change on the ground.

In practice, no company is singlehandedly working on the no-deforestation/SSCM/satellite monitoring nexus. There are supply chain collaborations, industry alliances, broader multi-stakeholder initiatives and work with technology providers. However, while bringing more actors on board means broader knowledge, expertise and perspectives, it can also require careful coordination and may redirect resources from action to coordination and slow down processes. What we found is that, considering the spectrum of starting positions, companies selectively decide who to partner with based on where their weaknesses or needs lie concerning the traceability, monitoring, follow-up and accountability functions (see Fig. 2) and build collaborations strategically to fill these gaps.

5 Challenges

This section dives into challenges and unknowns connected to satellite technology use in SSCM and for EUDR compliance.

5.1 Operational Challenges

Integration into SSCM practices remains a challenge and yet, effective satellite monitoring relies on a sound SSCM system. This includes for companies to understand their needs and to identify the right technology provider depending on their skill and expertise level. For instance, traders and integrated palm oil companies typically have GIS expertise in-house and may be well-served with just buying the data, whereas multinational brands that are far removed from raw material production typically need more support. Behind this operational challenge lies a communication challenge between corporate clients unfamiliar with space technology and technology providers unfamiliar with corporate SSCM. For technology experts to understand these nuances and the application context and scenarios faced by their clients (who may lack the technical understanding to even clearly articulate their needs or anticipate ethical concerns) is paramount and, probably, why we see more specialized technology providers in the supply chain sustainability space.

When it comes to **sustainability** more broadly, it is important to keep in mind that deforestation is only one aspect. There are numerous other environmental, labor or human rights aspects that go into full sustainability and should not be neglected—not the least because they tend to be closely connected to deforestation in the form of root causes or unintended side-effects. For instance, deforestation is often caused by rural poverty and lack of economic alternatives in smallholder communities which in turn increases the risk for child labor or poor working conditions on farms. Without accounting for these systemic linkages and addressing underlying issues, deforestation persists. Satellite monitoring is inherently not the ideal technology for monitoring most social sustainability aspects (think of human rights violations like modern slavery and forced labor or discrimination and harassment as well as labor rights or health impacts from exposure to pesticides) apart from perhaps monitoring and ensuring land rights. Yet, there is a range of providers out there that specialize in the labor and human rights side, e.g. Ulula's workers surveys and grievance mechanisms. Taking a more systemic approach and effectively integrating satellite technology with other elements is a skill in itself. The EUDR tries to address this limitation by incorporating the requirement for commodities to be produced in accordance with local laws, but this leaves companies with a patchwork of local regulations to consider while those regulations are often weak in the first place and unable to adequately protect workers, communities and indigenous peoples. Again, the EUDR is about raising the floor, not the bar, and leading companies have been going beyond these requirements for years—while realizing how difficult and complex it is to create lasting impact on the ground.

Along the same line, companies that adopt satellite technology for deforestation monitoring may discover **inconvenient truths** and issues that they did not know they had. This includes assuming that the supply chain is 'clean' to following up on a potential deforestation alert where earnestly investigating its root causes can easily uncover more systemic sustainability problems that are harder to deal with (see the previous paragraph). This comes back to the need for having a sound and comprehensive approach that includes preemptive procedures on how to react when a case of deforestation or child labor is detected and how to handle the supplier relation and further follow-up.

On this note, **interoperability** is an interesting challenge that will (hopefully) become even more relevant as companies are more tightly integrating different solutions for their SSCM. Interoperability of different solutions can occur either on the company level or on a solution or platform level. On a company level, if they want to integrate satellite data or solutions inhouse from different providers, e.g. working with one provider for deforestation monitoring and with another one for Lidar-based carbon stock accounting in a multi-stakeholder project, and perhaps also with e.g. internal procurement data and their analytics, this is more complex and requires more capabilities from the respective company. On the other hand, if different platforms or providers work on interoperability of their solutions or standardization of analysis ready data, this is ultimately more efficient and would benefit especially new clients entering the space.

5.2 Ethical and Information Security Concerns

5.2.1 Risk Avoidance on Paper vs. Sustainability Impact on the Ground

As raised in Sect. 4 in the context of follow-up and accountability, when using satellite monitoring, companies need to be prepared on how to deal with the insights gained and what the ethical implications are. A false dichotomy can emerge between avoiding risks on paper, i.e. (preventatively or reactively) terminating supplier relations to avoid association with deforestation, versus creating sustainability impact on the ground.

These dynamics is magnified by the EUDR. Overall, the law is ambitious and stringent, sending a clear signal that deforestation needs to end as a central part of climate change mitigation and pushing traceability and data-driven approaches. Simultaneously, it creates challenges. For instance, it is ambitious and strict—but is it missing the point? The stringent approach to ban any material at risk of being deforestation-linked may reduce 'imported deforestation' but it raises the question what the effect on overall deforestation rates will be. There is a risk that companies will preventatively pull out of higher risk regions and higher risk suppliers, e.g. smallholders where collecting geolocations and verifying legality involves more effort, and that producers and traders will simply redirect potentially deforestation-tainted material to non-EU markets. As companies scramble to set up their due diligence system, the EUDR's comparatively short implementation timeline and other regulation elements such as the benchmarking of countries into low-, medium- and high-risk for differentiated import checks risk fostering such short-term solutions. There is a risk that this may cause leakage and effect no change in deforestation rates on the ground but simply creates a two-pronged system of 'clean' produce flowing to the EU market while (possibly) deforestation-tainted material is redirect to other markets with lower sustainability standards. Similarly, the EUDR focuses on conducting due diligence to avoid association with sustainability issues, whereas most human rights due diligence approaches like the UN Guiding Principles for Business and Human Rights (where the concept originated for the supply chain context) emphasize conducting due diligence to detect, prevent and mitigate sustainability issues. The former incentivizes a 'clean vest' approach by companies whereas the latter emphasizes a 'getting your hands dirty, searching for problems and finding solutions' approach.

Speaking of effectiveness and potential side-effects, EUDR reviews and potential extensions are planned at specified intervals but risk irreversible and/or unintended side-effects until then. These reviews include the impact on 'Other wooded land' like the Cerrado or the Chaco and 'Other natural ecosystems' like mangroves or peatlands, which provide important ecosystem functions and are at risk of being converted to agricultural land. Including these areas could also provide important impulse for training algorithms to detect conversion in these non-forest areas. Other review topics include a potential inclusion of the financial sector in the EUDR due diligence requirements and potential impact on smallholders and indigenous peoples (scheduled for in five years).

5.2.2 Question of Justice and Smallholder Inclusion Under EUDR

I have already touched on this in other sections because smallholders play an essential and vulnerable role in many forest-risk commodities supply chains and are intertwined with other considerations. However, working with satellite imagery, Earth observation data and environmental sustainability issues more broadly, it is sometimes easy to forget that this work and these decisions have direct impacts on the lives of individuals and communities, thus I want to give this some additional space and also outline the upsides. Smallholders are already a vulnerable group in supply chains, but the EUDR risks aggravating this by creating incentives for companies to avoid sourcing from areas whose baseline deforestation/no-deforestation status is contested between different baseline maps, to preventatively stop sourcing from smallholders to avoid the effort of collecting geolocations and legality evidence, or alternatively, to push costs for mapping efforts onto smallholders without support, compensation or price premiums. In the human and labor rights space, recent worker-centric solutions allow companies to send surveys to workers who then get paid for their answers while retaining ownership of their data, e.g. Million Markers. Usually, workers participate without compensation while their answers eventually enable downstream buyers to boost reputation, attract investors and avoid regulatory trouble. Similar approaches for including smallholders on more equal terms and compensating them for provision of data (without transferring ownership) could fit into companies' EUDR due diligence process. Beyond this, there also lies potential in this for smallholders. Clarifying land rights and mapping farms can help smallholders get formal land titles. Satellite-based monitoring and EUDR compliance can help them verify deforestation-free status, access to the EU market and (possibly) better prices without the costly process of certification. Where buyers or traders invest in mapping farms and supplier development for compliance with local laws, this incentivizes companies to build longer-term relations with suppliers and pay premiums in exchange for continued access to this particular supply and provenance, a process also termed decommoditization.

On another level, the EUDR risks clashes with producing nations. This concerns first, the EU's overriding of local deforestation laws and deforestation definitions, and second, the EUDR's approach of benchmarking producer countries into high/standard/low risk with differentiated percentages of shipments to be checked by the Competent Authorities. Consequently, companies may reduce sourcing from high-risk countries. This turns the EU benchmarking exercise into a high-stake decision for producer countries' economies. Originally intended as an engagement tool for prioritizing producer countries for dialogue and support with EUDR implementation, this could have been more constructive, and possibly more effective overall. Still, there remains the continuous multi-stakeholder platform on forest protection and the 'Team Europe Initiative' for engagement with producer countries (European Commission, 2023a).

5.2.3 Data Integrity, Accuracy and Interpretation

At first glance, satellite imagery appears neutral and objective. When taking a closer look, however, it becomes clear that satellite data is not depicting one single, unambiguous truth but—depending on data quality and algorithms—can produce a variety of results. Taking satellite imagery at face value is at best naïve and at worst highly problematic. The same goes for 'deforestation' actually. While it seems a clear-cut term, multiple definitions exist and whereas the EUDR adopts the FAO's definition, producer countries may use others. After having clarified what is meant by deforestation, the question remains whether image quality allows accurate detection of said deforestation. Can it differentiate between deforestation, forest degradation, the extraction of individual high-value trees from otherwise intact natural forest, plantations of perennial crops or agroforestry? High quality imagery and combination of different spectral data sets can help get more accurate insights but require first, awareness of the limitations and second, ability to interpret and if needed integrate other information.

Similarly, the best data can become unreliable if the algorithms analyzing it are poorly trained or biased. Who developed and trains the algorithm? What is their expertise, background and (commercial) interest? What (peer) reviews have been undertaken? Also, more generally, there is the question of data integrity. Especially concerning due diligence processes for EUDR compliance, it is important to ensure the accuracy and completeness of information used, i.e. making sure that the imagery has not been altered, corrupted or even willingly falsified to remove evidence of deforestation.

The current discussion around EUDR baseline maps illustrates these challenges. The 2020 forest cover baseline map by the EU Forest Observatory (2023) has been criticized as unreliable and overstating forested areas. Satellite monitoring providers with longstanding expertise (Satelligence, 2024) have pointed out that the baseline map falsely identifies smallholder farms and perennial crops like cocoa or oil palm as forest. This raises questions on how to proceed and what forest cover map to rely on. Most technology providers have proprietary baseline maps. However, while perhaps more accurate, these maps are not public and have not undergone the wide peer review that is currently enabling criticism of the EU Forest Observatory's map. Further, it is unclear how the Competent Authorities will check for deforestation and which baseline map they will rely on. What is the point of using a more accurate map if its data is not accepted in official checks? Further, producer countries have national forest baseline maps but since their deforestation definitions may differ from the EU's, these are not applicable either. These uncertainties may further lead companies to adopt conservative baseline estimates to avoid shipments from being held up and preventatively stop sourcing from smallholders or specific regions. In this sense, overstating data's 'objectiveness' can quickly (and/or inadvertently) create or reinforce injustices and discrimination against vulnerable groups in supply chains. As adoptions rises among companies, regulatory authorities and NGOs that engage with satellite technology for the first time, they may overestimate data's accuracy especially under tight EUDR timelines, making technology providers important partners for building awareness and understanding of satellite monitoring's limitations.

5.2.4 Data Ownership, Storage and Sharing

Effective use of satellite monitoring requires traceability, and thus data sharing, across the supply chain. As discussed before, gaining visibility into the supply chain is difficult because suppliers tend to be hesitant to share information on who they are sourcing from. This information can be commercially sensitive, constituting their competitive advantage, or risking that buyers use information to drive down prices or leapfrog them. Similarly, data sharing across the supply chain can clash with anti-trust rules. For these reasons, there is a need for data sharing agreements and standards and clarify beforehand questions such as who owns the data, how is the data used, shared, stored and protected etc.

Now, these aforementioned challenges exist concerning any supply chain (sustainability) data sharing. What is particular about deforestation monitoring and the EUDR is that it involves geolocation data that can spill over into personal information, as well as comprehensive data on legality in terms of compliance with local labor, tax, anti-corruption etc. laws. As an example, quite practically, what does free, prior and informed consent (FPIC) look like for indigenous peoples and smallholder farmers under EUDR? Considering that information on geolocations and legality needs to be in place by the end of 2024, and companies may need time to even trace material back, this leaves limited time for FPIC processes. And how free is consent if access to the EU market—and possibly one's livelihood—depends on it? Whereas personal information of EU citizens is protected and regulated under the EU General Data Protection Regulation (GDPR), it remains unclear how data e.g. geolocations of smallholders farms (and often homes) is protected. Similarly, viewing data as a form of currency, ethical questions arise around who provides data and who benefits from it. Approaches could be transferred from the human and labor rights space where workers are compensated for providing data while retaining ownership (see Sect. 5.2.2).

On another note, public concerns around misuse of satellites for spying on individuals have been increasing e.g. around Albedo's very-low orbit imaging satellites with 10-cm resolution that are primarily serving defense and intelligence purposes and have sparked concern about privacy issues and ethics of satellite monitoring (Broad, 2024). While not necessarily directly linked to corporate deforestation monitoring, it is relevant to reflect on where technological possibilities are heading.

These questions are not straightforward to answer and may be messy to address but they need to be asked. Satellite monitoring and sustainability data sharing are useful tools but to enable effective and responsible use, we need awareness and discussions around the thornier parts.

6 Conclusion and Outlook

The momentum in space technology is strong and there lies tremendous potential in applying satellite monitoring for sustainable sourcing and building transparent global supply chains.

This chapter traces back why and how companies have been starting to adopt and experiment with satellite monitoring of their supply chains' sustainability issues (especially deforestation). We look at the background of how supply chain management and Earth observation work (Sect. 2), and explore how they fit together in the realm of satellite monitoring for sustainable sourcing by tracing underlying developments that drive adoption (Sect. 3): Companies face increasingly serious business repercussions from sustainability issues deep in their supply chains, e.g. deforestation on palm oil plantations, yet lack the visibility and oversight over what is happening on the ground. Coinciding with immense technological progress, regulatory pressure from the EU Deforestation Regulation and hopes for more data-driven decision-making, satellite monitoring is thus on the rise. The chapter takes stock of current (best) practices and deep-dives into the practical nitty–gritty of implementation. While satellite technology has tremendous potential to enable a more data-driven approach in supply chain sustainability, it is critical to keep in mind that it is no silver bullet but requires careful integration with sound sustainable supply chain management and necessary collaborations to effectively create on-the-ground change. It is **not** a quick fix, but requires companies to have traceability, follow-up capacities and stakeholder accountability, plus external partnerships where needed, to be actionable—all of which can take years to build (Sect. 4). The chapter then wraps up with a broader discussion of operational challenges, sustainability dilemmas and ethical concerns that (may) arise when adopting satellite monitoring with the ambition to address root causes of deforestation and create lasting change towards sustainability. These challenges and the importance of addressing them are amplified through the EUDR (Sect. 5).

Overall, the EUDR is sending a strong signal and if anything gives an immense push for satellite monitoring and data-driven approaches towards traceability and supply chain sustainability. As a broader outlook, satellite monitoring has tremendous potential for supply chain sustainability issues: Beyond forest cover monitoring, the continuous advances in spectral resolutions and algorithms now allow space-based monitoring of carbon stocks and emissions, and other planetary variables, e.g. by Planet, enabling (corporate) climate and biodiversity action. Similarly, integration with other digital solutions such as workers survey tools, e.g. Ulula's solutions, allows to capture and integrate human and labor rights aspects for systemic solutions.

References

Akhavan, R. M., & Beckmann, M. (2017). A configuration of sustainable sourcing and supply management strategies. *Journal of Purchasing and Supply Management, 23*(2), 137–151. https://doi.org/10.1016/j.pursup.2016.07.006

Beske, P., Land, A., & Seuring, S. (2014). Sustainable supply chain management practices and dynamic capabilities in the food industry: A critical analysis of the literature. *International Journal of Production Economics, 152*, 131–143. https://doi.org/10.1016/j.ijpe.2013.12.026

Broad, W. J. (2024). When eyes in the sky start looking right at you. *NY Times*, 1–12. https://www.nytimes.com/2024/02/20/science/satellites-albedo-privacy.html?utm_source=substack&utm_medium=email1/12

Choi, T. Y., & Wu, Z. (2009). Taking the leap from dyads to triads. Buyer-supplier relationships in supply networks. *Journal of Purchasing and Supply Management*, *15*(4), 263–266. https://doi.org/10.1016/j.pursup.2009.08.003

Csillik, O., Kumar, P., Mascaro, J., O'Shea, T., & Asner, G. P. (2019). Monitoring tropical forest carbon stocks and emissions using Planet satellite data. *Scientific Reports, 9*(1), 1–12. https://doi.org/10.1038/s41598-019-54386-6

Dubois, A. (2009). Comment on "Taking the leap from dyads to triads: Buyer-supplier relationships in supply networks" by Choi and Wu. To leap or not to leap: Triads as arbitrary subsets of networks of connected dyads. *Journal of Purchasing and Supply Management*, *15*(4), 267–268. https://doi.org/10.1016/j.pursup.2009.08.002

EU Forest Observatory. (2023). *Global map of forest cover for year 2020*. https://forest-observatory.ec.europa.eu/forest/gfc2020

European Commission. (2023a). *Deforestation regulation implementation*. https://green-business.ec.europa.eu/deforestation-regulation-implementation_en

European Commission. (2023b). *FAQs EU Deforestation Regulation*. https://circabc.europa.eu/ui/group/34861680-e799-4d7c-bbad-da83c45da458/library/e126f816-844b-41a9-89ef-cb2a33b6aa56/details

European Commission. (2023c). *Regulation (EU) 2023/1115 of the European Parliament and of the Council of 31 May 2023 (EUDR)*.

Foerstl, K., Reuter, C., Hartmann, E., & Blome, C. (2010). Managing supplier sustainability risks in a dynamically changing environment-Sustainable supplier management in the chemical industry. *Journal of Purchasing and Supply Management, 16*(2), 118–130. https://doi.org/10.1016/j.pursup.2010.03.011

Ghadge, A., Wurtmann, H., & Seuring, S. (2020). Managing climate change risks in global supply chains. A review and research agenda. *International Journal of Production Research, 58*(1), 44–64. https://doi.org/10.1080/00207543.2019.1629670

Godar, J., Suavet, C., Gardner, T. A., Dawkins, E., & Meyfroidt, P. (2016). Balancing detail and scale in assessing transparency to improve the governance of agricultural commodity supply chains. *Environmental Research Letters*. https://doi.org/10.1088/1748-9326/11/3/035015

Gong, Y., Jiang, Y., & Jia, F. (2023). Multiple multi-tier sustainable supply chain management: A social system theory perspective. *International Journal of Production Research, 61*(14), 4684–4701. https://doi.org/10.1080/00207543.2021.1930238

Halldórsson, Á., Kotzab, H., Mikkola, J. H., & Skjøtt-Larsen, T. (2007). Complementary theories to supply chain management. *Supply Chain Management: An International Journal, 12*(4), 284–296. https://doi.org/10.1108/13598540710759808

Hansen, M. C., Potapov, P. V., Moore, R., Hancher, M., Turubanova, S. A., Tyukavina, A., Thau, D., Stehman, S. V., Goetz, S. J., Loveland, T. R., Kommareddy, A., Egorov, A., Chini, L., Justice, C. O., & Townshend, J. R. G. (2013). High-resolution global maps of 21st-century forest cover change. *Science, 342*(6160), 850–853. https://doi.org/10.1126/science.1244693

Heldt, L. (2020). Sustainability risks in multi-tier supply chains. A dynamic capabilities perspective on integrating remote sensing technology into sustainable sourcing to manage sustainability risks in upstream supply chains (IIIEE Master Thesis).

Heldt, L., & Beske-Janssen, P. (2023). Solutions from space? A dynamic capabilities perspective on the growing use of satellite technology for managing sustainability in multi-tier supply chains. *International Journal of Production Economics, 260*. https://doi.org/10.1016/j.ijpe.2023.108864

Hofmann, H., Busse, C., Bode, C., & Henke, M. (2014). Sustainability-related supply chain risks. Conceptualization and management. *Business Strategy and the Environment, 23*, 160–172. https://doi.org/10.1002/bse.1778

Hofmann, H., Schleper, M. C., & Blome, C. (2018). Conflict minerals and supply chain due diligence. An exploratory study of multi-tier supply chains. *Journal of Business Ethics, 147*(1), 115–141. https://doi.org/10.1007/s10551-015-2963-z

Jamalnia, A., Gong, Y., Govindan, K., Bourlakis, M., & Mangla, S. K. (2023). A decision support system for selection and risk management of sustainability governance approaches in multi-tier supply chain. *International Journal of Production Economics, 108960*. https://doi.org/10.1016/j.ijpe.2023.108960

Lopatin, E., Trishkin, M., & Gavrilova, O. (2016). Assessment of compliance with PEFC forest certification indicators with remote sensing. *Forests, 7*(4), 1–12. https://doi.org/10.3390/f7040085

Lozano, R. (2015). A holistic perspective on corporate sustainability drivers. *Corporate Social Responsibility and Environmental Management, 22*(1), 32–44. https://doi.org/10.1002/csr.1325

Marin-Burgos, V., Clancy, J. S., & Lovett, J. C. (2015). Contesting legitimacy of voluntary sustainability certification schemes: Valuation languages and power asymmetries in the roundtable on sustainable palm oil in Colombia. *Ecological Economics, 117*, 303–313. https://doi.org/10.1016/j.ecolecon.2014.04.011

Mena, C., Humphries, A., & Choi, T. Y. (2013). Toward a theory of multi-tier supply chain management. *Journal of Supply Chain Management, 49*(2), 58–77. https://doi.org/10.1111/jscm.12003

Partzsch, L., & Vlaskamp, M. C. (2016). Mandatory due diligence for 'conflict minerals' and illegally logged timber: Emergence and cascade of a new norm on foreign accountability. *Extractive Industries and Society, 3*(4), 978–986. https://doi.org/10.1016/j.exis.2016.07.003

Planet. (2020). *Planet tasking. Sub-meter, sub-daily tasking.* https://www.planet.com/products/hi-res-monitoring/

Roberts, S. (2003). Understanding the Patchy success of ethical sourcing initatives. *Journal of Business Ethics, 44*, 159–170.

Satelligence. (2024). *Making open forest data safe for EUDR compliance.* https://satelligence.com/making-open-forest-data-safe-for-eudr-compliance/

Schepers, D. H. (2010). Challenges to legitimacy at the Forest Stewardship Council. *Journal of Business Ethics, 92*(2), 279–290. https://doi.org/10.1007/s10551-009-0154-5

Seuring, S., & Müller, M. (2008). From a literature review to a conceptual framework for sustainable supply chain management. *Journal of Cleaner Production, 16*(15), 1699–1710. https://doi.org/10.1016/j.jclepro.2008.04.020

Sisodiya, N., Dube, N., & Thakkar, P. (2020). Next-generation artificial intelligence techniques for satellite data processing. *Remote Sensing and Digital Image Processing*. https://doi.org/10.1007/978-3-030-24178-0_11

Smit, H., Mcnally, R., & Gijsenbergh, A. (2015). *Implementing deforestation-free supply chains. Certification and beyond* (Issue February). http://theredddesk.org/resources/implementing-deforestation-free-supply-chains-certification-and-beyond

Sodhi, M. M. S., & Tang, C. S. (2019). Research opportunities in supply chain transparency. *Production and Operations Management, 28*(12), 2946–2959. https://doi.org/10.1111/poms.13115

Tachizawa, E. M., & Wong, C. Y. (2014). Towards a theory of multi-tier sustainable supply chains: A systematic literature review. *Supply Chain Management: An International Journal, 19*(July), 643–653. https://doi.org/10.1108/SCM-02-2014-0070

Tropical Forest Alliance. (2017). The role of the financial sector in deforestation-free supply chains. In *World Economic Forum* (Issue January).

Turner, W., Spector, S., Gardiner, N., Fladeland, M., Sterling, E., & Steininger, M. (2003). Remote sensing for biodiversity science and conservation. *Trends in Ecology and Evolution, 18*(6), 306–314. https://doi.org/10.1016/S0169-5347(03)00070-3

Werner, T. T., Bebbington, A., & Gregory, G. (2019). Assessing impacts of mining: Recent contributions from GIS and remote sensing. *The Extractive Industries and Society, 6*(3), 993–1012. https://doi.org/10.1016/j.exis.2019.06.011

Managing the Orbital Debris Hazard Long-Term: Space Sustainability

Darren McKnight

Abstract The orbital debris environment, especially in low Earth orbit (LEO), is dynamic and concerning. The accumulation of abandoned massive derelicts (e.g., rocket bodies and non-operational payloads), clouds of fragments from breakup events, and thousands of new smallsats, often deployed in large constellations, creates challenges for achieving space sustainability. The interdependency of these three main constituents of LEO provides a compelling means to examine emerging collision risks. Technical terms and equations used to quantify collision risk and prioritize risk reduction measures are provided. The probability of collision for specific encounters for operational spacecraft are provided by Conjunction Data Messages (CDMs) as the key currency for space traffic coordination. The annual probability of collision to operational spacecraft from clouds of fragments depicts the primary statistical collision risk posed to spacecraft. The collision rate within clusters of massive derelicts abandoned decades ago is shown to characterize the primary debris-generating potential in LEO.

Keywords Orbital debris · Space safety · Space traffic coordination · Debris mitigation · Space sustainability · Debris remediation

1 Introduction

Orbital debris is any manmade object in Earth orbit that performs no useful purpose. This may include abandoned rocket bodies, non-operational spacecraft, and remnants from the fragmentation of space hardware. Unfortunately, any orbit where humans have operated has been sullied by orbital rubbish being left behind; even the Moon has been littered by manned spaceflight and robotic missions. However, in Earth orbit the altitudes below 2,000 km, called low Earth orbit (LEO), has had the most amount of activity and thus the greatest number of orbital debris left behind.

D. McKnight (✉)
LeoLabs, Menlo Park, CA, USA
e-mail: darren@leolabs.space

H. Jahankhani et al. (eds.), *Space Law Principles and Sustainable Measures*, Space Law and Policy, https://doi.org/10.1007/978-3-031-64045-2_7

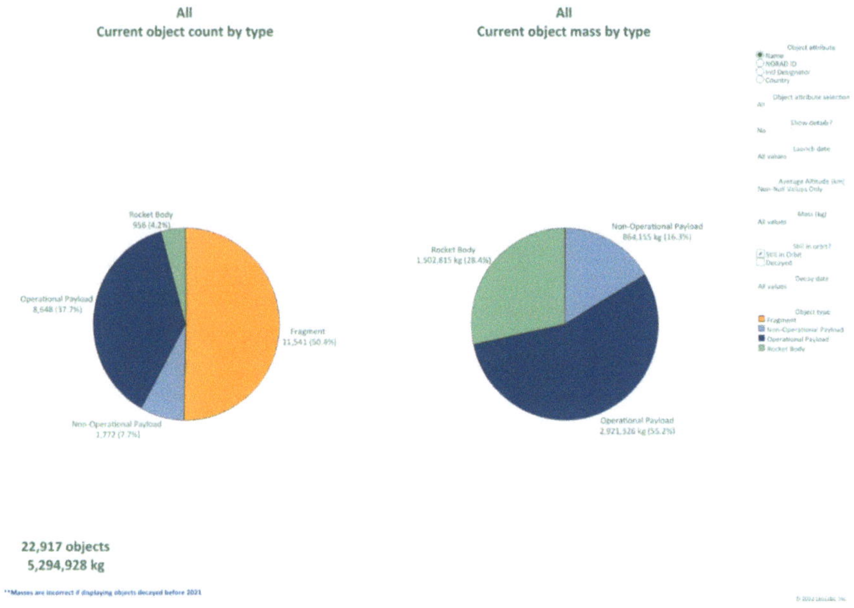

Fig. 1 About half of the LEO population is comprised of fragments from satellite breakups

While operational payloads are certainly an important component of the LEO population, even with a plethora of launches, ~60% of objects in the LEO population are comprised of space debris (i.e., derelict intact hardware, such as rocket bodies and nonoperational payloads, combined with fragments from hundreds of satellite fragmentations). This orbital debris also comprises ~40% of the total mass in LEO.

In summary, the LEO population consists of clusters (of massive derelicts), clouds (of fragments), and constellations (of operational payloads); each exhibit different and interesting collision risk patterns. Figure 1 shows the distribution in number and mass in LEO as of 12 July 2024.

2 The Life of a Rocket Body

The collision risk in low Earth orbit (LEO) can be a confusing mix of terms and equations. In order to better understand the importance of collision risk for the short-term safety of operational constellations and long-term sustainability of LEO, we will follow an individual space object through a sequence of risk calculations. This process will clarify and highlight the need for continuous monitoring and characterizing of all components of the LEO space population to catalyze actions to control the growth of debris.

2.1 Key Characters in Our Story

On May 22, 1990, the Soviet Union launched Cosmos 2082, Tselina-2, an Electronics Intelligence (ELINT) satellite with a mass of 3,250 kg, into an 840 km circular orbit (Fig. 2).

Before examining the three major constituents of the space object population in LEO, the "life" of a single massive derelict is examined to introduce some useful terms and equations.

The Zenit-2 launch vehicle (also called SL-16) deployed this satellite and the upper stage from this launch vehicle was left in a similar orbit to Cosmos 2082. Cosmos 2082 is also denoted as 1990-46A (i.e., the primary object of the 46th launch of 1990) and Satellite Number 20624 (i.e., the 20624th object cataloged by the now 18th Space Defense Squadron of the now United States Space Force). The SL-16 rocket body (R/B) has a mass of 9,000 kg, is roughly 11 m in length, and has a diameter of 4 m. (Fig. 3) The SL-16 R/B is designated 1990-46B and Satellite Number 20625.

The practice of leaving a R/B along with a deployed payload was used by many space operators in the 1980s and 1990s. Both of these objects have now been in orbit about the Earth for over 32 years, circling the globe over 165,000 times. The distance these two objects have traveled is just under eight billion km; this is equivalent to a round trip from the Sun to Pluto and back!

On 5 June 2022, the S-16 R/B 20625 was involved in a close approach with another derelict rocket body, the 37 kg Scout X-4 upper stage, deposited in LEO in 1964 by the United States to deploy Explorer 25 (INJUN-4). The Conjunction Analysis Report, provided by LeoLabs, details this conjunction of 483 m miss distance with

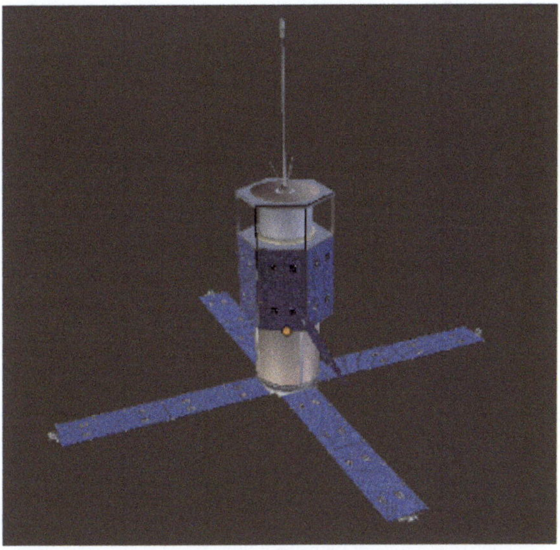

Fig. 2 Tselina-2 from Tselina-2 | 3D Resources (nasa.gov)

a relative velocity of 8.7 km/s (i.e., ~17,000 mph) with a probability of collision (PC) of 1.3E-3 (i.e., ~1/800 chance). While this may seem like a low probability, typically an operational spacecraft in LEO will take evasive actions if an encounter has a PC greater than 1E-5 (i.e., a hundred times less likely than the close approach that occurred). In 2023, there were over 37,000 close approaches in LEO where the PC exceeded 1E-5. The Conjunction Analysis Report summarizes what is included in a Conjunction Data Message (CDM) (Fig. 4).

Figure 5 depicts how the CDM provides the PC between two distinct objects for one specific encounter.

In 2023, object 20625 had seven conjunction events with a PC over 1E-5, shown in Fig. 6.

That large number of high-PC conjunctions involving 20625 in 2023 is only the tip of the iceberg. In 2023, there were 58 conjunctions with PC greater than 1E-6 for this specific SL-16 R/B. The distribution of the objects 20625 encountered in orbit

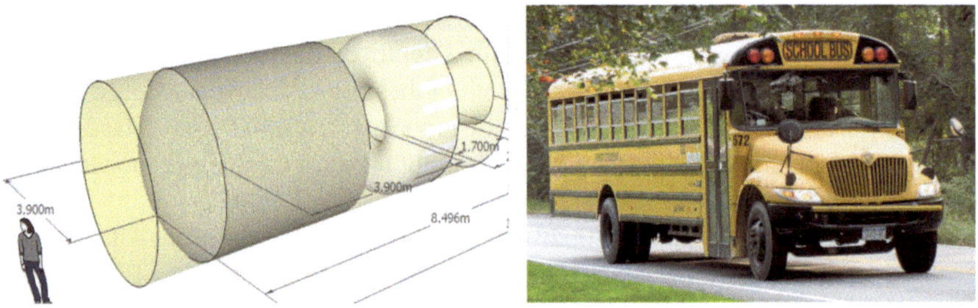

Fig. 3 The SL-16 rocket body is similar mass and dimensions to a big yellow school bus except it is going 7.5 km/s, has no brakes, and no steering wheel

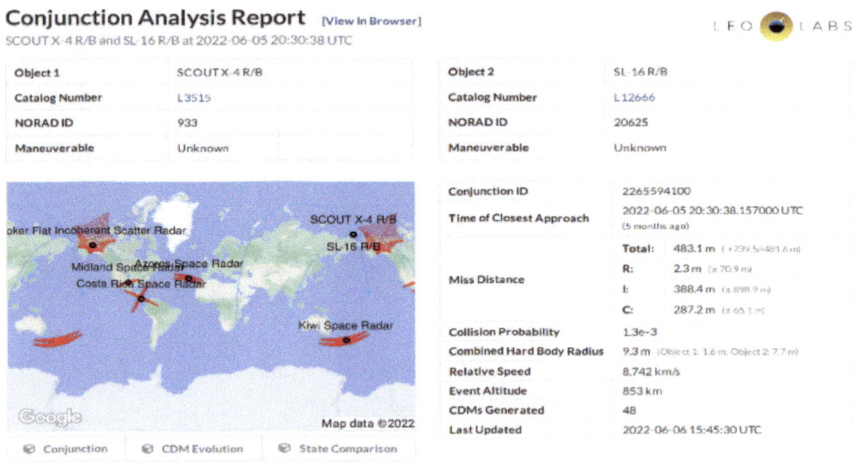

Fig. 4 The conjunction analysis report for a conjunction summarizes the salient features of the close approach between two space objects

Fig. 5 Probability of collision between two objects for a single close approach is given in a CDM

	Object 1	Object 2	TCA	Miss Distance	PoC	Relative Speed	CDMs
View	SL-16 R/B Catalog Number: L12666 NORAD ID: 20625	CZ-6A DEB Catalog Number: L134790 NORAD ID: 55894	2023-11-26 01:34:53 UTC (3 months ago)	4.241 km	1.5e-5	11.162 km/s	56
View	SL-16 R/B Catalog Number: L12666 NORAD ID: 20625	CZ-6A DEB Catalog Number: L133241 NORAD ID: 54617	2023-10-19 19:33:01 UTC (4 months ago)	0.477 km	1.2e-5	5.150 km/s	9
View	SL-16 R/B Catalog Number: L12666 NORAD ID: 20625	ONEWEB-0631 Catalog Number: L133834 NORAD ID: 55171	2023-10-07 02:20:02 UTC (5 months ago)	3.294 km	3.1e-5	14.605 km/s	19
View	SL-16 R/B Catalog Number: L12666 NORAD ID: 20625	CZ-6A DEB Catalog Number: L134784 NORAD ID: 55886	2023-04-11 05:33:51 UTC (10 months ago)	6.647 km	1.1e-5	11.545 km/s	87
View	SL-16 R/B Catalog Number: L12666 NORAD ID: 20625	NOAA 16 DEB Catalog Number: L15025 NORAD ID: 41523	2023-03-02 21:06:55 UTC (a year ago)	1.837 km	6.3e-5	6.591 km/s	44
View	SL-16 R/B Catalog Number: L12666 NORAD ID: 20625	COSMOS 1408 DEB Catalog Number: L125892 NORAD ID: 49528	2023-02-04 03:00:10 UTC (a year ago)	8.600 km	1.9e-5	5.323 km/s	10
View	SL-16 R/B Catalog Number: L12666 NORAD ID: 20625	CZ-6A DEB Catalog Number: L133026 NORAD ID: 54432	2023-01-25 06:11:41 UTC (a year ago)	0.613 km	6.0e-5	13.838 km/s	13

Fig. 6 Satellite 20625 was involved in seven conjunctions in 2023 with the PC exceeding 1E-5

from 1 January 2022 to 12 July 2024 is shown in Fig. 7 produced by the LeoLabs' LeoMap tool.

While ~96% of the encounters are with fragments there were 12 events with operational payloads and three with other R/Bs. This examination hints at the larger perspective of all objects in the vicinity of our SL-16 R/B 20625 posing a collision risk to this massive derelict.

2.2 *Statistical Probability of Collision (PC)*

The statistical PC for a single object from all of the resident space objects that might cross the object's orbit is determined using an equation that considers the object's size (i.e., collision cross-section) times the spatial density of other objects (i.e., number of objects per cubic kilometer) times the relative velocity between these objects (~12 km/s in LEO) (Fig. 8).[1]

The LeoLabs LeoRisk tool provides a way to determine the statistical annual PC levels and what objects contribute to the collisional hazard for any object in LEO. As shown in the figure below, the types of objects 20625 encounters throughout the

[1] PC = $1\text{-}e^{\lambda t}$ where λ = collision cross-section (km^2)*relative velocity (km/s)*spatial density (#/km^3) and t is time (s).

Fig. 7 Depiction of all encounters with 20625 over two years shows that fragments are the most likely conjuncting object

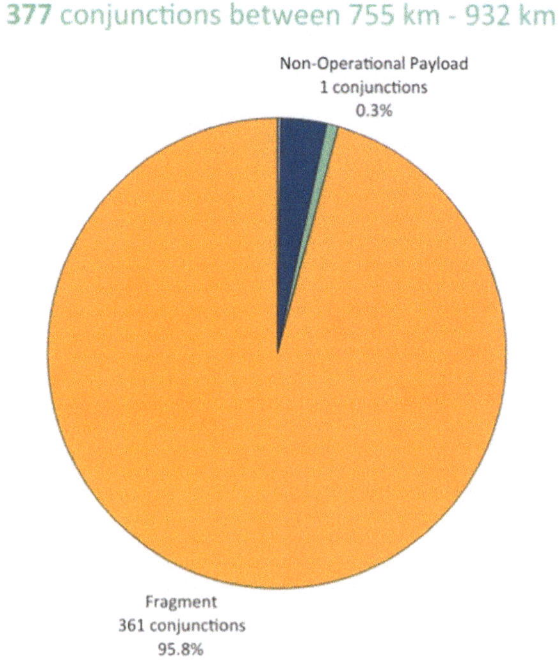

Fig. 8 Probability of collision for one year provides a statistical measure of collision likelihood

course of a year to reach a PC of 4.51E-3 (i.e., ~1/220) is within a factor of eight of the summation of CDMs issued during 2023. Note, the summation of the CDM PC values will not necessarily be equal to the statistical PC level over a short period such as a year; it may take a decade or more for these values to line up, if at all (Fig. 9).

The consequence of these events with 20625 may range from minor to very consequential. For example, a fragment impacting 20625 may create hundreds of pieces of debris while a collision between 20625 and the Tselina-2 satellite (20624) that it deployed would likely create over 15,000 cataloged fragments and many more lethal nontrackable debris. Now, let's examine how likely it is for these most massive objects (i.e., SL-16 R/Bs and non-operational Tselina-2 satellites) to collide in LEO.

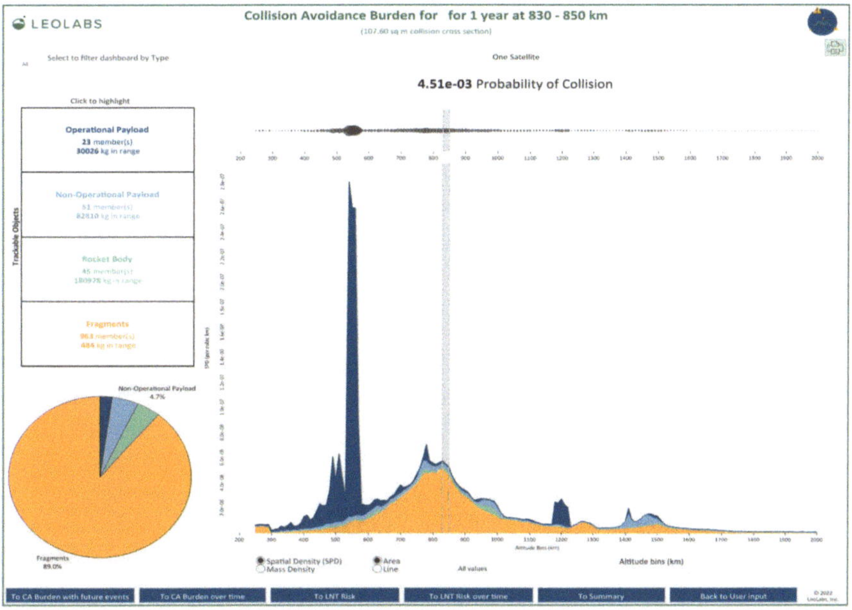

Fig. 9 The statistical PC over a year for 20625 is 4.51E-3 (i.e., ~ 1/220)

2.3 *Cluster 840 as a Hotspot in LEO*

The original characters in this risk story, 20624 and 20625, were part of a larger deployment by the Soviets/Russians. In total, from 1985 to 2007, 18 ELINT satellites and their associated 18 SL-16 upper stages were deposited between 825 and 865 km. While our previous collision hazard calculations examined a single event between two objects against each other (i.e., a CDM) and an annual risk for one object vs all other space objects it could possibly encounter (i.e., statistical PC), a collision rate (CR) can be determined among any subset of objects. In this case, that subset is these 36 massive objects deposited centered around 840 km that we call Cluster 840 (C840). This dynamic is depicted in Fig. 10.

These 36 objects alone amount to 208,000 kg, this is equivalent to over ~800 Starlink V1 satellites or ~1,400 OneWeb satellites (i.e., about double of the eventual

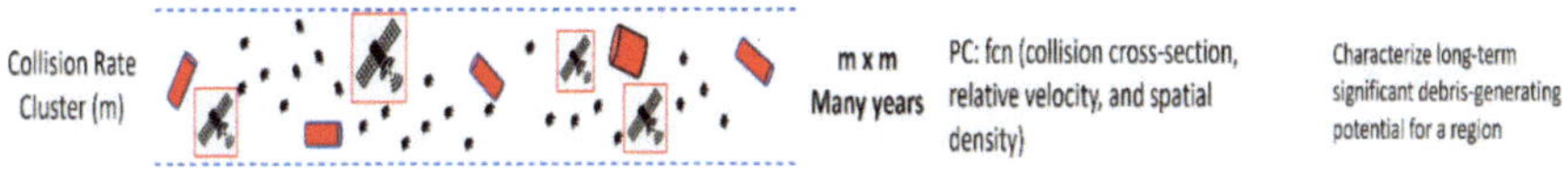

Fig. 10 Collision rate within a subset of the space object population is the summation of PCs of any two objects within the population

constellation deployment plans). CR is a relevant concept[2] when any collision event amongst a population is meaningful. In this case, the catastrophic collision between any two objects in C840 would likely create over 15,000 cataloged fragments, in essence doubling the LEO fragment population in one instance. This debris would likely be spread over many hundreds of kilometers. For example, the debris cloud from the fragmentation of the Chinese Fengyun-1C satellite in 2007 at 860 km have been involved in over 19,000 high-PC conjunctions at altitudes as low as ~330 km and as high as ~1,630 km in 2023 alone. It is also important to note that debris created at these altitudes will linger for many decades, even centuries.

The current annual collision rate within C840 is 1.7E-3 (i.e., a 1/580 chance each year).[3] The importance of C840, however, is amplified when we remember this ensemble of derelict mass has been whizzing past each other for decades (i.e., the Sun to Pluto and back). Taking this into account, the probability that the first collision between these objects may occur by 2025 is ~6%. By 2040, this expectation will increase to ~9% without any intervention such as removing some of these objects. The 18 SL-16 R/Bs in C840 have been highlighted on numerous occasions as posing a significant debris-generating potential in LEO and have been identified as 18 of the top 50 objects that should be removed from LEO.[4]

3 Other Clusters of Massive Derelicts

Many of the most massive objects added to the LEO population were abandoned in the last 20 years. The littering of LEO with defunct rocket bodies is a long sad consequential common practice. The seventh space launch ever—deployment of Vanguard 2 in 1959—left a rocket body that is still in LEO today, more than 64 years later! Since 2004, ~282 more R/Bs were abandoned in LEO (above 500 km average altitude) with an average mass of over 2,000 kg (i.e., over 150 times more massive than a 6U cubesat). These R/Bs were left by launches from seven different countries—including every major spacefaring entity. Nearly 40% (i.e., 132) of these R/Bs were left by China (two within the "bad neighborhood" of 800 to 900 km), while Russia only left ~50 R/Bs in LEO in the last 20 years. It should be noted there were an additional ~30 rocket bodies deposited in higher orbits outside of LEO (i.e., typically much more long-lived than LEO) in 2023. The "bad neighborhood" (which

[2] CR = $(N^2/2)$*collision cross-section (km^2)*relative velocity (km/s)*relative velocity (km/s)*time (sec)/Volume of altitude expanse of cluster (km^3) where N is the number of objects in the cluster.

[3] The CR is for any of the 36 massive derelicts colliding with each other which would result in nearly 15,000 cataloged fragments while the annual PC for 20,625 involves any object interacting with 20,625; this includes hundreds of fragments.

[4] McKnight, et al., "Identifying the 50 Statistically-Most-Concerning Derelict Objects in LEO," 71st International Astronautical Congress (IAC)—The CyberSpace Edition, Dubai, UAE, October 20 and McKnight, D., Dale, E., Bhatia, R., Kunstadter, C., Stevenson, M., and Patel, M., "A Map of the Statistical Collision Risk in Leo", 73rd International Astronautical Congress, Paris, France, September 2022.

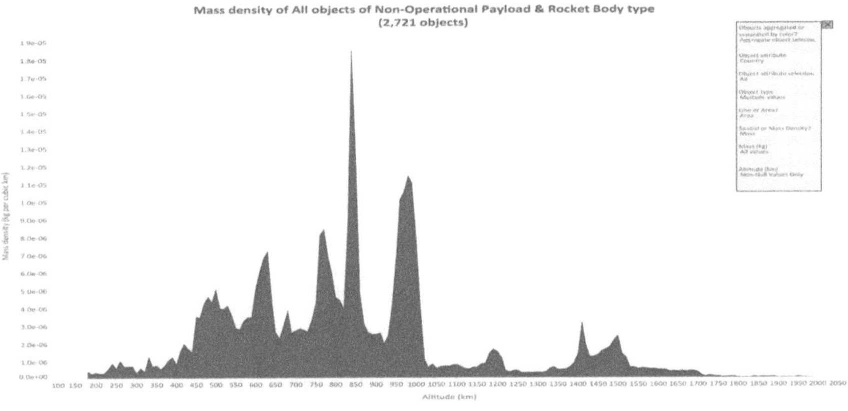

Fig. 11 The mass distribution of derelict objects provides a mapping of clusters

includes C840), noted above and in previous reports by LeoLabs, is where many of the most massive derelicts have been abandoned historically. This region continues to be a hot spot for debris collision risk. In 2023, there were 1,352 conjunctions in LEO with a miss distance of less than 100 m. Over 20% of these most dangerous conjunctions (i.e., 291 instances) occurred between 800 and 900 km. Plotting the mass density of derelict objects in LEO provides a mapping of all the clusters at 840 km, 980 km, 770 km, and 630 km (in decreasing order). Figure 11 shows the accumulation of mass by altitude.

4 Clouds of Fragments

The second major constituent of LEO, after clusters of massive derelict objects, are clouds of fragments.

As shown earlier, ~50% of the current LEO satellite catalog is comprised of fragments (i.e., ~11,541 out of ~22,917 as of 15 February 2024). While most fragments are the majority by number, the mass that they represent is almost trivial in comparison to the ~11,000 intact objects in LEO. Fragments are usually much smaller than a meter in size with typical "fragments" being on the order of 10 to 50 cm.

The term "debris" is used by the 18th Space Defense Squadron (SDS) whenever there is an object related to a launch that is neither a payload nor a rocket body. This "debris" may include a small number of mission-related items such as payload adapters, lens covers, retaining clamps, etc. There have been ~310 launches that have only one piece of debris associated with it and another ~130 with only two pieces associated with the launch. Summing up all the debris that include single digit numbers associated for a given launch amounts to ~1,400 objects (i.e., 12%) of the ~11,500 "fragment" population.

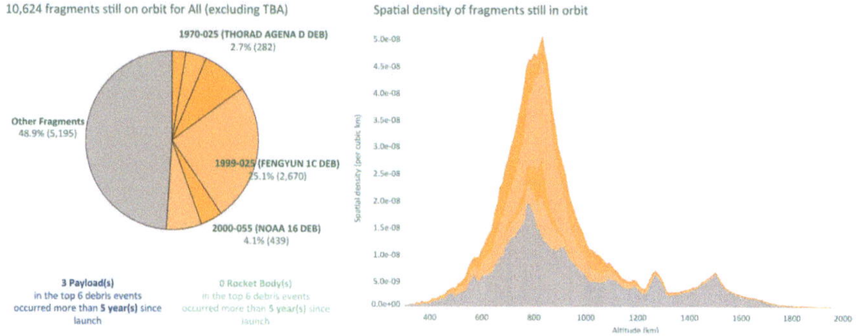

Intl Designator Group	*Cause*	*Event Parent Type*	*Breakup year*	*Years since launch*	*Breakup altitude (km)*	*Max Pieces*	*Distinct count of Still in Orbit Debris*	*Percent still in orbit*
1999-025 (FENGYUN 1C DEB)	Deliberate Collision	Payload	2007	8	855	3,519	2,670	75.9%
1993-036 (COSMOS 2251 DEB)	Accidental Collision	Payload	2009	16	775	1,711	914	53.4%
2022-151 (CZ-6A DEB)	Accidental Explosion	Rocket Body	2022	0	840	793	710	89.5%
1981-053 (COSMOS 1275 DEB)	Accidental Explosion	Payload	1981	0	980	475	414	87.2%
2000-055 (NOAA 16 DEB)	Accidental Explosion	Payload	2015	15	850	458	439	95.9%
1970-025 (THORAD AGENA D DEB)	Unknown	Rocket Body	1970	0	1075	437	282	64.5%

Fig. 12 Half of all fragments on orbit in LEO came from just six breakup events

At the other end of the spectrum, major breakup events have liberated hundreds, even thousands, of fragments from a single mission. More specifically, the top six breakup events, in terms of the remaining debris on orbit, contribute 50% (i.e., ~5,200 fragments) of the current fragment population in LEO. Further, the top 24 breakup events encompass a full 75% of the total fragment population in LEO. The fragment population peaks at ~840 km but has two smaller peaks at ~1,250 km and ~1,500 km; at these altitudes, this debris will not reenter for centuries. The source of these fragments is primarily rocket bodies, however, four out of the top six most dominant clouds have payloads as their parent (Fig. 12).

It is very important to note that the three major space powers each have two of the top six most prolific breakup events. Accidental explosion is the most frequent cause for these breakup sources; amongst the top six events three of them were accidental explosion and the Thor Agena D R/B event is listed as unknown but is likely also an accidental explosion. These fragments pose a significant collision risk across LEO: from 1 January 2022 to 15 February 2024, there were ~808,000 conjunctions in LEO where the probability of collision (PC) exceeded one in a million (i.e., 1E-6). Of these events, ~385,000 (i.e., 48%) involved at least one fragment.

Looking more closely at the risk posed by fragment clouds, where risk is defined by PC times the mass involved in each conjunction, a different aspect of the collision hazard is exposed. Figure 13 plots the aggregate risk (i.e., debris-generating potential) from each cloud on the x-axis and the total number of conjunctions in which it was involved on the y-axis. The NOAA 16 fragment cloud surpassed the Cosmos 2251 cloud in aggregate risk, despite having fewer fragments on orbit, due to its proximity to the most massive, abandoned derelicts in LEO in the cluster centered around 840 km.

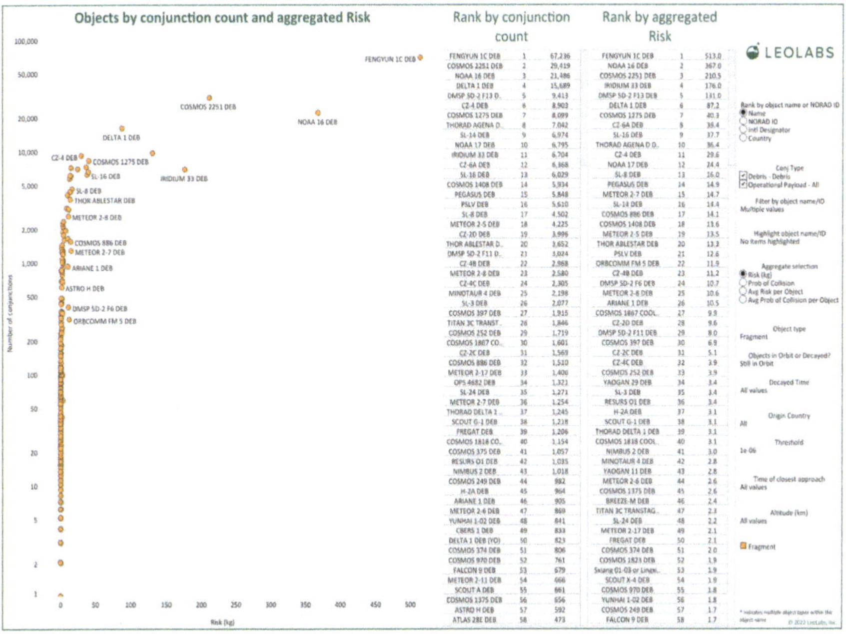

Fig. 13 Examining the aggregate risk (i.e., summing up PC times mass involved in each conjunction) posed by each debris cloud shows that NOAA 16 rises above Cosmos 2251 due to the NOAA 16 fragment cloud being centered within C840 which has many of the most massive derelicts in all of LEO

It is also important to note that 12 of the top 24 major fragmentation events occurred more than five years after deployment. This means that if the new 5-year post mission disposal (PMD) guideline had been applied a significant amount of the current fragment burden in LEO could have been avoided. Whereas, only three events occurred more than 25 years after the parent object was launched into space. This means that even if the old 25-year PMD rule had been adhered to, it would not have prevented much of the current fragment population from being generated. This reminds us all that the sooner derelict objects can be removed from orbit, the less likelihood that significant long-lived, debris-generating events will litter the LEO environment. While only one of these 24 events involved an object launched in the last 20 years, the frequency of breakup events over time has stayed fairly constant at about two to four per year. The altitude of these major events is typically over ~750 km since atmospheric drag will act quickly on fragments below this altitude; there have been many more breakup events that have occurred below ~750 km but they have not had persistent effects on the LEO population.

It is noted that the most long-lived debris clouds are the ones involving payloads versus rocket bodies. This is partially due to the more compact construction of payloads. Rocket body debris may contain a significant amount of thin-walled structure representing the roughly cylindrical shape of rocket bodies while payloads will contain dense components such as batteries, structural members, and payload hardware (e.g., transmitters, cameras, etc.).

In examining the contributions by country to the population of fragments in LEO, ~90% of the population originated from the US, China, and Russia. The peak of the Chinese contribution is clearly from the Fengyun-1C ASAT Test conducted in 2007 while the contribution from Russia (peaking at ~750 km) is largely from the remnants of Cosmos 2251 that collided with the American Iridium-33 satellite in 2009. However, the US contribution peaks just above ~800 km from a number of DMSP/NOAA spacecraft "minor" fragmentation events that add up to a significant contribution to the fragment collision hazard in LEO.

The beginning of 2022 saw the quick rise of cataloged fragments from the purposeful destructive on-orbit antisatellite (ASAT) test by Russia of their Cosmos 1408 (C1408) payload. The fragmentation event eventually consisted of over ~1,800 fragments cataloged but never more than ~1,200 fragments in the catalog at any one time. The ~1,200 fragment peak occurred in March 2022 has dwindled to ~65 by 15 February 2024.

Interestingly, despite the rapid decline of the number of C1408 fragments in the catalog, the number of Conjunction Data Messages (CDMs) issued involving a fragment from the C1408 breakup did not follow the same steep drop-off. This was partially due to the increased number of operational satellites deployed in 2022 that were characteristically challenged by C1408 fragments in the 400 km to 600 km altitude range.

However, just as the C1408 fragment cloud was being cleansed from LEO, the explosion of a Chinese CZ-6A rocket body in the middle of C840 (but, of course!) on 12 November 2022, partially reversed this trend of a shrinking fragment count. One day after the CZ-6A rocket body deployed the Yunhai 3 weather satellite into an 854/856 km orbit, the rocket body exploded. The total number of cataloged fragments from this event is 793 of which 710 (i.e., ~90%) are still in orbit (Fig. 14).

While this event occurred in the most densely populated region in LEO, since it occurred so soon after deployment, it was likely triggered by something related to the propulsion system and is likely not a collision-induced fragmentation. It may have been triggered (1) by attempted venting of remaining propellants, (2) if an orbit-lowering burn was attempted, or (3) the rocket stage simply failed to shut down smoothly.

The potential of it being propulsion-related is reinforced by reporting of Cees Bassa that "observations from two consecutive passes over the US in the hours after launch show fuel leaking from the rocket" as filmed by Dan Bush of Missouri Skies. Satellite Fuel Dump ??? Possibly Chinese CZ-6A launch—YouTube (https://www.youtube.com/watch?v=mG2kMff6vAM) Other energy sources on the rocket body also likely include batteries and pressurized vessels related to the propulsion system.

A fragment cloud can be represented in two ways: a Gabbard diagram and a spike plot. The Gabbard diagram for the CZ-6A rocket body shows the distribution of objects in LEO from as low as ~320 km and to as high as ~1,500 km.

The LeoLabs' LeoBreakup tool has assessed the CZ-6A R/B event as a high intensity explosion largely because (1) some of the objects with the greatest impulse

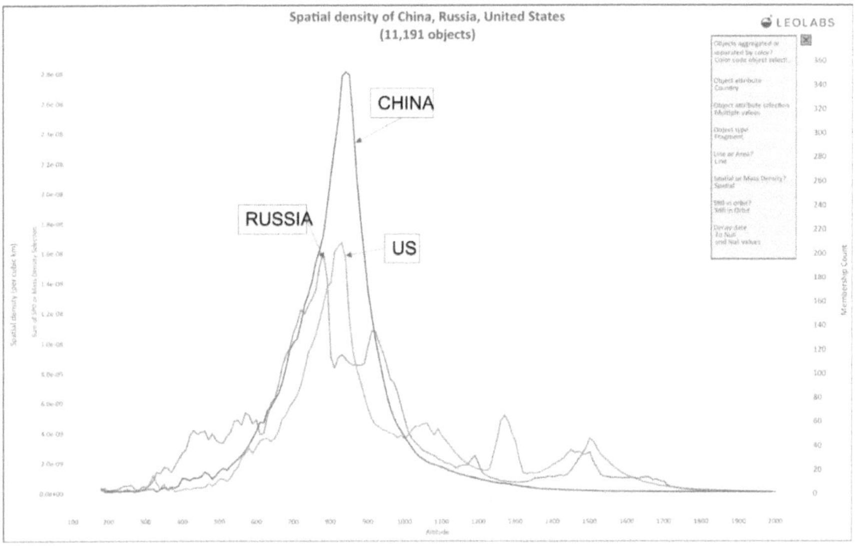

Fig. 14 The three major space powers have contributed significantly to the on-orbit fragment population

imparted to them (i.e., moved to the highest orbits) were some of the largest objects, (2) the number of cataloged objects is small in comparison to the mass of the rocket body (i.e., many fewer cataloged fragments than the mass of the rocket body in kilograms), and (3) it occurred so soon after payload deployment (Fig. 15).

The spike plot provides an immediate measure of the collision risk posed to other resident space objects from a breakup event by plotting the spatial density of the fragment cloud as a function of altitude. The peak and average spatial density values can be compared to the existing background population levels at these altitudes. The spatial density at 830 km before the event was 5.6E-8, where the fragment cloud from the CZ-6A rocket body peaks at 5.0E-9. As a result, the collision probability at 830 km has risen ~9% as the result of this breakup. This percentage increase drops off rapidly to under 2% below 700 km and above 900 km.

In 2023, there were nearly 6,000 conjunctions with a PC greater than 1E-6 involving a fragment from this event. These events occurred between 355 km to 1,601 km in altitude, as shown in the figure below. Further, the number of conjunctions with a PC greater than 1E-5 and 1E-4 were 666 and 84, respectively. The figure from LeoLabs' LeoMap tool shows the distribution of conjunction events as a function of time and type of object with which the CZ-6 fragments are conjuncting; over half of the events involved an operational payload and ~11% were with rocket bodies. Of the ~700 events involving rocket bodies, 158 were with the massive SL-16 rocket bodies, proposed as the most important derelict objects to remove from LEO (Figs. 16 and 17).

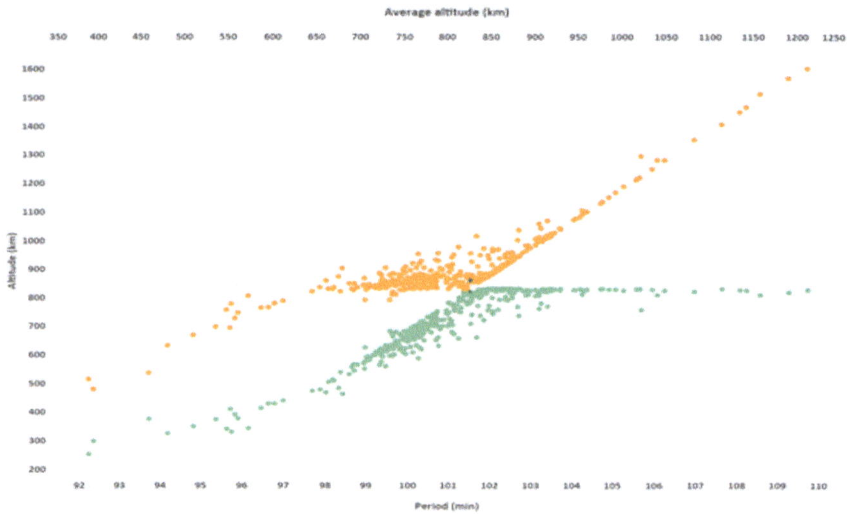

Fig. 15 A fragment cloud is often represented with a Gabbard diagram showing the distribution of fragments by orbital period (*x*-axis) and altitude extremes (*y*-axis)

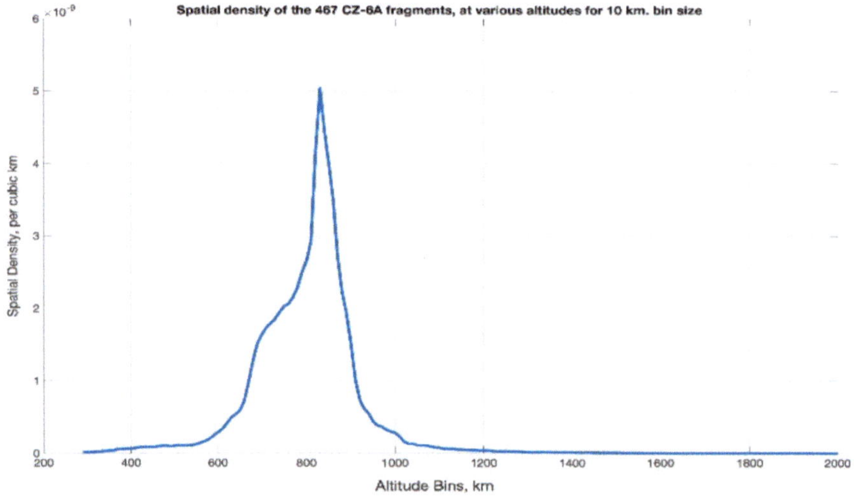

Fig. 16 A spike plot, created by LeoLabs for measuring effects of a fragment cloud on the on-orbit population, plots the spatial density (i.e., # of objects per cubic kilometer) against altitude

5 Constellations of Operational Payloads

While Starlink has captured the imagination and the high-level statistics regarding the absolute number of payloads added to their constellation (~1,600 in 2022 and ~1,950 in 2023), taking a closer look shows several other constellations also grew

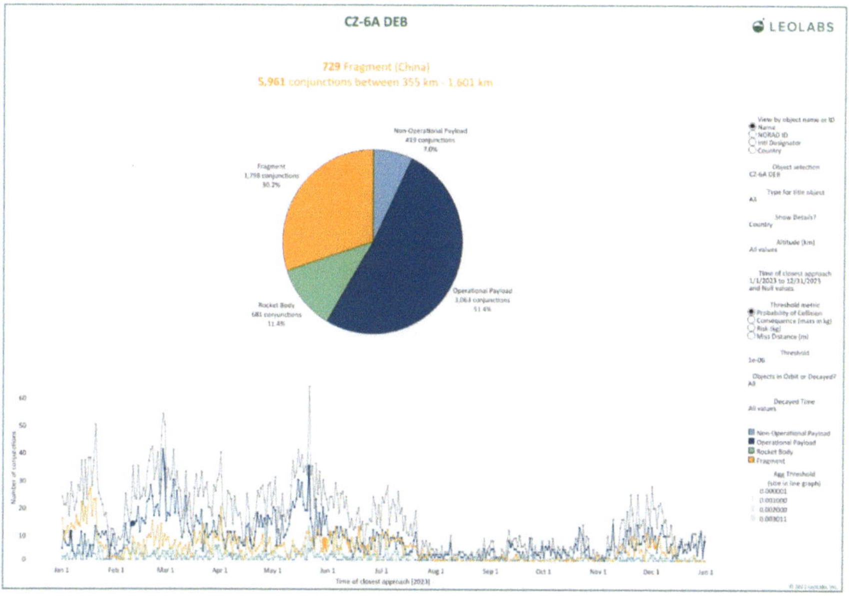

Fig. 17 The fragment cloud from the recent CZ-6A R/B explosion has contributed significantly to the collision risk in LEO over the last 15 months

significantly since 2021: OneWeb, Planet, SpaceBee, and Spire Global all had double digit percentage growth rates of their constellations over this time.

As the number of operational payloads continues to increase, the importance of space traffic coordination grows. This is reflected in the changing proportion of high probability conjunctions (i.e., PC > 1E-6), with an increase in what we call "space traffic management (STM)" conjunctions relative to "space debris management (SDM)" conjunctions. A STM conjunction is any close approach that includes at least one operational payload while a SDM event is a conjunction between two debris objects (i.e., neither object is an operational payload). Since 2021, the balance of high-PC conjunction events involving an operational payload has gone from being 40% of the events to being 95% of events by the beginning of 2024.

Examining the filings for satellite constellations in the US and news reports abroad it is seen that over 95% of operational satellites in LEO by 2030 will be from six constellations: SpaceX V1 & V2, Guowang, Kuiper, G60, and OneWeb. Figure 18 depicts the likely trend in operational satellites from these six constellations alone through 2030.

While this growth will require significant care by spacecraft operators, current constellations have shown the capability to operate thousands of satellites in close proximity very safely and reliably. Of course, we cannot ignore the space safety risk from the "other" 2,000 to 6,000 operational satellites (i.e., not in large constellations). Some satellite operators have less robust risk reduction processes so in aggregate tens or hundreds of satellites may pose a greater collision risk than thousands of

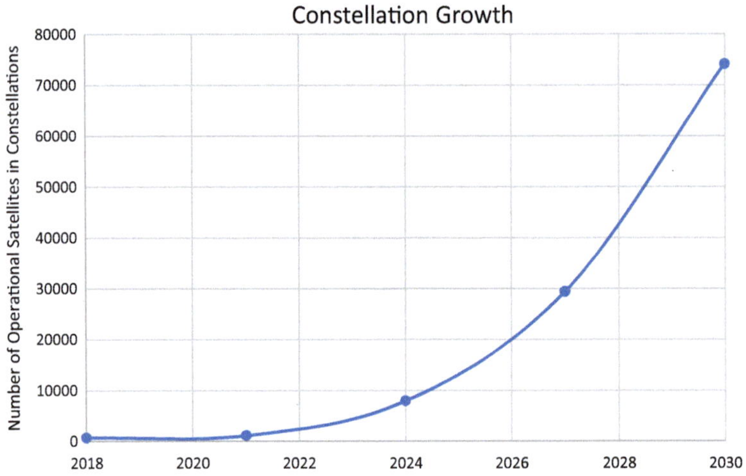

Fig. 18 The growth of satellites to be part of major constellations will continue to rise exponentially through 2030 challenging space situational awareness and space traffic coordination resources and processes

responsibly-operated satellites. Further, the continued generation of thousands of fragments and abandonment of hundreds of massive derelicts will continue to present an operational collision risk to even the most responsible satellite operators.

It should be noted that the catalog population only represents about 10% of the lethal debris population in LEO. There is an estimated 250,000 fragments between 1 and 10 cm that are not currently being tracked by any entity to provide sufficient information to enable risk reduction maneuvers (RRMs) to avoid all of the lethal collision hazard. This hazard increases exponentially as altitude increases since the atmospheric drag at lower altitudes accelerates the reentry of the all objects but especially fragments, as they have characteristically higher area-to-mass ratios.

6 Three Ways to Manage the Orbital Debris Environment

The characteristics of the three major constituents highlight how much of LEO is accumulating numbers and mass of debris that are mathematically a ticking time bomb that the global aerospace community must cooperatively work to manage.

The three primary means to control the debris collision risk and support space sustainability are:

- Conjunction Data Messages (CDMs) are the currency for timely and effective **space traffic management**. Massive derelicts (i.e., rocket bodies and non-operational payloads) and fragments cannot execute risk reduction maneuvers but most of the over ~8,400 operational payloads in LEO can act on these warnings.

- Statistical probability of collision highlights the need to stop adding debris to LEO (i.e., **debris mitigation**).
- The collision rate between clusters of massive derelicts amplifies the need for **debris remediation** (i.e., active debris removal).

It is hoped that the continual discussion of these facts and figures will help motivate policymakers, regulators, space operators, and international military services to act responsibly to preserve the space environment for the generations to come. As a member of the next generation—Tamara from Williamston High School in Michigan—stated so eloquently in a recent letter to LeoLabs: "The longer that the government waits to help solve this problem, the more debris there will be in orbit."

The persistence of the massive derelicts in the clusters centered around 840 km and 980 km bodes poorly when considering the multi-decade length of these objects interacting. As a matter of fact, at these two altitudes the probability that there will be a collision between the massive derelicts by 2025 are ~6% and ~18%, respectively.

While constellations with literally thousands of members are getting the lion's share of the attention it is critical for all satellite operators to maintain high standards of space safety practices to include robust/responsive propulsion system, consistent/transparent PC thresholds for maneuver/abatement, sharing propagated ephemeris, and open communications.

Massive derelicts continue to be abandoned in long-lived (i.e., ≥25 yr) orbits, since 2004 over ~300,000 kg in ~140 R/Bs have been abandoned in LEO. This adds to the already significant debris-generating potential brewing at high altitudes within LEO with little hope for immediate removal missions over the next few years. Some good news is that fragmentation events have been abating slightly in number and severity over the last 10 years as is evidenced that only one of "top six" breakups occurred in the last decade. However, the rapid increase in operational satellites coupled with the increasing debris-generating potential of the massive derelicts and dispersed fragment clouds lingering in LEO paints a concerning picture as the global space economy is rumbling forward at a rapid pace.

7 Parting Comments

All the action "in" LEO of the last few years has not been in orbit; three major policy advancements of note have matured space safety principles and provided some positive trends to watch carefully to offset the ever-increasing space safety and space sustainability issues.

First, the U.S. Federal Communications Commission (FCC) implemented a reduction of the 25-yr rule (i.e., require licensees to show that their payloads will be removed from orbit within 25 years after the end of their mission lifetime) to a 5-yr rule. This forward-leaning change is indicative of national and international entities understanding the importance of leaving less debris in orbit from operational missions is important moving forward for enhancing enduring space safety (i.e., space sustainability).

Second, the United States spearheaded, through the United Nations, a call for a global moratorium on destructive on-orbit anti-satellite (ASAT) testing. This initiative is reflective of the enormous contribution of the last two on-orbit ASAT tests has had on collision risk in LEO; ~10% of all high-risk conjunctions in LEO since 1 January 2022 involved a fragment from one of these two events.

Third, Senator Hickenlooper's (Democratic Senator from Colorado) staff rallied a bipartisan initiative that culminated in the Senate unanimously passing the Orbital Sustainability Act of 2022 (i.e., ORBITS Act). The ORBITS Act is focused on operationalizing active debris removal on massive derelict objects that have been left in orbit by previous U.S. Government space missions. It is exciting to see the U.S. commit to remediate some of the mass that poses such a great debris-generating potential.

Deepfakes in Social Engineering Attacks

Darren Steve Dsouza, Ayman El Hajjar, and Hamid Jahankhani

Abstract This research delves extensively into deepfake technology's substantial influence on cybersecurity, revealing its subtle aspects and potential social repercussions. The investigation begins with deciphering the practical process for generating deepfakes using DeepFace Lab, providing light on every tedious step from data extraction to model training and video conversion. The study delves into the domain of social engineering attacks, explicitly depicting situations in which hostile actors use deepfakes to influence trust. In such instances, the combination of realistic visual and audio aspects raises the sophistication of cyber threats, needing more awareness and strong authentication protocols. The research delves further into the dark side of political manipulation, giving a clear picture of how perfectly constructed deepfakes may destabilise political landscapes. The hypothetical situation provided highlights the potential harm deepfakes can have on public trust and democratic processes, as well as the difficulty established media sources confront in real-time verification. The research closes by advocating for an initiative-taking and collaborative reaction by experts, legislators, and society to strengthen defences against the weaponization of fabricated media, emphasising the fragile equilibrium between technical advancement and the preservation of democratic principles.

Keywords Deepfake · Social engineering attacks · AI · Cyber threats · Cybersecurity · DeepFaceLab

D. S. Dsouza · A. E. Hajjar
University of Westminster, London, UK

H. Jahankhani (✉)
Northumbria University, London, UK
e-mail: Hamid.jahankhani@northubria.ac.uk

H. Jahankhani et al. (eds.), *Space Law Principles and Sustainable Measures*, Space Law and Policy, https://doi.org/10.1007/978-3-031-64045-2_8

1 Introduction

The rapid advancements in digital technology have ushered in an era of unprecedented change, with over half of the developing world's population connected to the digital realm in a trivial two decades (United Nations, 2020). The digital realm is currently experiencing a shift towards the Fifth Industrial Revolution (5IR) dubbed the AI revolution. This transition is facilitated by the widespread use of 4IR technologies including Artificial Intelligence (AI), Big Data (BD), and the Internet of Things, among others (Chang & Park, 2020; Golić, 2020; Whittaker et al., 2023). This revolution represents a unique convergence of human ingenuity and machine capability, resulting in the production of goods at an unprecedented rate (Golić, 2020).

The impact of AI is ubiquitous, extending across all sectors of society. Projections indicate that the AI industry, which was estimated to be worth USD 136.6 billion in 2022, is poised for remarkable growth, with an expected average annual compound growth rate of 37.3% from 2023 to 2030 (Grand View Research, 2022). In the financial sector alone, AI has demonstrated its capacity to enhance customer service (76%), boost operational efficiency (75%), and reduce costs (74%) (Microsoft, 2018). AI has unquestionably offered countless advantages and considerably enhanced numerous facets of human life.

However, it is critical to recognise that AI, like any other technical progress, is not without its drawbacks. 'Every coin has two sides,' as the expression goes, and AI is no exception. One such adverse consequence is the emergence of deepfakes. Deepfakes encompass various forms of manipulated media content, including videos, audio recordings, and other formats, deliberately crafted to present a distorted depiction of reality (van der Sloot & Wagensveld, 2022).

This research aims to look at deepfake technology's practical applications, ethical concerns, and cybersecurity consequences. It delves into the DeepFaceLab development process, emphasising ethical considerations and societal implications. The research dives into the potential risks brought about by deepfakes in social engineering attacks on IT security, emphasising the necessity for strong cybersecurity measures. It also investigates the use of deepfakes in political manipulation, notably during elections, highlighting the threats to democratic processes. The ultimate objective is to provide nuanced insights that enlighten politicians, cybersecurity professionals, and the general public on the risks associated with deepfake technology while pushing for initiative-taking efforts to solve these concerns.

The journey into the area of cybersecurity has been an ongoing interest, fuelled by the constant news stories of individuals falling victim to social engineering attacks (SEA). What is fascinating is the attack's significant dependence on human psychology rather than technical expertise. Individuals with little to no technical knowledge might plan successful attacks by exploiting human weaknesses.

The experiences in SEA are more than just academic; it is deeply rooted in personal experiences that have left an indelible mark. There have been several phishing attempts on the author, particularly on social media sites. A friend once reached out on Instagram, unable to access their account, and requested the code Instagram had sent for login verification. This episode served as a clear reminder of the devious tactics used in these attacks as the attacker would gain access to the account using the code.

Furthermore, SEA has been witnessed on dating apps, where conversations are steered away from the platform to other social media, like Instagram. Conversations build trust over time, eventually leading to video call requests which upon further investigation, was discovered that these tactics were part of a broader scheme involving extortion.

The evolution of these schemes, combined with emerging technologies like deepfakes, blurs the line between reality and fiction, making such hoaxes appear more trustworthy and capable of causing severe harm.

2 Literature Review

The rapid rise of digital technology has resulted in tremendous breakthroughs and risks in cybersecurity, among which the rise of deepfakes has had a significant impact on the field of social engineering attacks. The resemblance of these synthetic content to reality makes them challenging to identify. They are frequently produced using sophisticated artificial intelligence algorithms to alter audio, video, and images. This analysis of the literature aims to identify gaps in the existing knowledge and provide a comprehensive assessment of the impact of deepfakes on attacks using social engineering in cybersecurity. The review seeks to broaden the conversation on cybersecurity by introducing diverse viewpoints and laying the groundwork for future studies that address these new cyber threats.

2.1 Deepfakes History and Overview

The manipulation of images and videos isn't novel; software like Photoshop allows for altering images that range from minor adjustments to substantial modifications (Jones, 2013). With the proliferation of machine learning tools, the sophistication of doctored media has significantly advanced (Cassauwers, 2019). Deepfakes, as outlined in Westerlund (2019), Kietzmann et al. (2019), and Mirsky and Lee (2021), represent AI-driven manipulations of media content, leveraging deep learning algorithms. These techniques proficiently substitute one individual's likeness with another's, encompassing entirely fabricated or partially manipulated media, such as videos and audio, often with deceptive intent (Albahar & Almalki, 2019; van der Sloot & Wagensveld, 2022). The term "deepfakes" was coined by an anonymous Reddit user who blended "deep learning" with "fakes" (Kietzmann et al., 2019). In late 2017, this user shared a video on the platform, wherein Gal Gadot's countenance was artfully superimposed onto an existing explicit video, crafting a convincing fake (Department of Homeland Security, 2020). Furthermore, the user also divulged the underlying code facilitating the creation of manipulated videos, leading to a proliferation of fabricated content (Kietzmann et al., 2019). This phenomenon led to a rapid upsurge in user-generated manipulated explicit content, with tens of thousands of people partaking in the dissemination of such content on the platform (Maddocks, 2020).

2.2 *Deepfake Generation Techniques*

Deepfake creation is a novel method of media manipulation that overcomes the limits of traditional imitation techniques. It significantly reduces recognisable traces or patterns commonly used to detect forgeries, such as discrepancies in biometric data or compression artefacts. Prior research has extensively highlighted the efficacy of deepfakes in reducing these identifiable markers commonly used in detecting manipulated content (Galbally et al., 2014; Kanwal et al., 2019; Verdoliva, 2020). This advancement signifies a departure from traditional forgery detection methods, enabling the creation of manipulated media that evades common detection mechanisms. Manipulations that do not require sophisticated artificial intelligence (AI) tools are sometimes referred to as "shallow fakes" or "cheap fakes" (Verdoliva, 2020). Showing a small portion of a longer video to give the impression that it is the full thing is a classic instance of a cheap fake. In an instance with Joe Biden discussing Ukraine, a segment was edited to suggest the US President speaking incoherently (Reuters Fact Check, 2023). The full video provides the context that the edited clip lacks.

Deepfakes employ deep neural networks to understand and extract input characteristics, enabling the reconstruction of entirely new, hyperrealistic content based on learned segmentation maps or latent representations derived from the input data (Seow et al., 2022). Seow et al. (2022) broadly classify Deepfake generation into three models:

- Autoregressive model
- Autoencoder
- Generative Adversarial Network (GAN)
- Autoregressive Models.

Autoregressive models, known for their ability to generate realistic images, play a crucial role in deepfake creation by capturing extended dependencies within images, particularly in rendering realistic, lifelike facial images (Aaron et al., 2018; Chen et al., 2018; Oord et al., 2016). Chen et al. (2018) elaborates on the utilization of recurrent neural networks (RNNs) by these models to represent the conditional distribution of the subsequent element, relying on past elements. The efficacy of this model is associated with its capability to manage long-range dependencies, frequently mitigated through the application of causal convolutions. However, challenges in computational demands and data prerequisites arise when utilizing these models for deepfake generation (Aaron et al., 2018; Oord et al., 2016). Pixel-CNNs (Van den Oord et al., 2016) serve as an example of autoregressive models.

- Autoencoder

The autoencoder, an artificial neural network, focuses on unsupervised data representation learning, involving an encoder and decoder network to transform input data into a latent representation and reconverting it back to its original form (Kingma & Welling, 2019; Li et al., 2023). Noteworthy in deepfake creation, the Variational Autoencoder (VAE) differs from standard autoencoders by assuming a specific probability distribution of latent variables, enhancing complex data restoration (Li et al., 2023). Despite potential blurriness in VAE's outputs, it excels in generating new outputs post-training by sampling the distribution (Li et al., 2023).

- Generative Adversarial Network (GAN)

GAN, comprising a generator and discriminator network, operates by generating synthetic outputs to deceive the discriminator and aims to discern real from fake samples (Nguyen et al., 2022). Through continual optimization, both networks reach an equilibrium where the fake data becomes indistinguishable from real data. This method supports challenging manipulations such as style transfer and image restoration, not easily achievable through traditional methods (Chen et al., 2020). Over time, various GAN iterations have enhanced software performance (Karras et al., 2017; Rao & Yang, 2020; Seow et al., 2022). Examples like StyleGAN (Tewari et al., 2020) and FaceApp (deepfakes, 2023) excel in entertainment-based deepfake video production, while the RCNN network (Dong et al., 2016) bolsters mobile camera resolution. Stacked-GAN addresses low-quality deepfake synthesis, employing super-resolution to retain more facial details (Chen et al., 2022). However, GAN necessitates substantial computational power and extensive training datasets (Karras et al., 2017).

2.3 *Applications of Deepfake Technology*

Deepfake technology offers a plethora of constructive applications. For instance, it enables seamless voiceovers in foreign movies (Zeitchik, 2021), enriches learning by recreating ancient figures for a more interactive educational experience (Silbey & Hartzog, 2019), promises to potentially reinstate speech for individuals affected by conditions such as ALS, granting the use of the individual's original voice (Chesney & Citron, 2018), and facilitates virtual dress-up during shopping (Rincon, 2023). While deepfake technology offers a multitude of constructive applications, it's essential to recognize that these sophisticated manipulations also find a darker purpose in the realm of cybersecurity, particularly within Social Engineering Attacks (SEAs), which will be detailed in the subsequent section.

2.4 Social Engineering Attacks

Social engineering, as defined by Mitnick and Simon (2002), involves manipulating individuals, primarily through psychological persuasion, to illicitly access systems and acquire confidential data. Despite advancements in security measures safeguarding sensitive information, the human element persists as a vulnerable link due to susceptibility to manipulation (Mouton et al., 2016). As highlighted by Krombholz et al. (2015), social engineers exploit human vulnerability, convincing individuals to disclose sensitive information or carry out malicious actions through persuasion. Despite the general belief in individuals' ability to detect such attacks, research by Qin and Burgoon (2007) and Marett et al. (2004) indicates their poor performance in discerning lies and deception. These attacks encompass physical, social, and technical elements, employed at various stages of the attack (Krombholz et al., 2015).

- Instances

Flick and Morehouse (2011) present several instances of Social Engineering Attacks (SEAs):

- Impersonating an employee to alter their password via the IT Help Desk.
- Posing as service vendors (e.g., document shredder service, maintenance workers) to acquire sensitive data or sabotage equipment.
- Placing USB drives in targeted areas, like the headquarters' parking lot, containing malware to access IT infrastructure.
- Sending "phishing" emails to extract sensitive information or IT infrastructure details from the company's employees.

Furthermore, these attacks are intrinsically linked to data breaches, intending to illicitly access data. Mitnick and Simon (2002), emphasize this aspect, highlighting the overarching goal of social engineering attacks. Notably, according to the FBI's report in 2022, 82% of breaches resulted from human error, with a significant 34% increase in reported phishing-related complaints. The financial ramifications of these breaches are substantial. In the United States, the average cost of a data breach soared to $9.44 million in 2022, up from $9.05 million in 2021, with a global average cost of $4.35 million per breach in 2022 (Statista, 2023). Illustrative cases like Kevin Mitnick's infamous attacks underscore the severe impact of sophisticated social engineering attacks on both corporate and governmental information security (Mitnick & Simon, 2002). The success of social engineering does not necessarily demand extensive technical expertise; rather, it preys on common human psychological traits such as curiosity, courtesy, gullibility, greed, apathy, diffidence, and thoughtlessness (BBC News, 2012).

- Techniques

Social Engineering Attacks can be broadly classified into human-based or computer-based. Until 2006, psychological manipulation formed the core of social engineering attacks (Aldawood & Skinner, 2018; Peltier, 2006). Peltier (2006) expanded this definition to include technology-based tactics in these crimes. Human-centric attacks involve person-to-person strategies like impersonation or persuasion, while technology-based methods encompass phishing, vishing, and other online scams (Aldawood & Skinner, 2018; Salahdine & Kaabouch, 2019). Social engineering attacks employ diverse techniques to extract sensitive information, yet the fundamental pattern of the attack remains consistent. According to Salahdine and Kaabouch (2019), SEAs traditionally follow a template consisting of four steps:

1. Conduct target research.
2. Form a connection with the object of attack.
3. Exploitation of the relationship and organising an attack and
4. Exit and leave no evidence.

In the initial phase of a social engineering attack, gathering information about the target through physical means is crucial. The success of subsequent phases and the overall outcome heavily rely on the data obtained during this phase (Gallegos-Segovia et al., 2017). Targets are selected using straightforward search methods, which may involve public documents and face-to-face interactions (Aldawood & Skinner, 2020).

Following the research phase, the primary focus lies on nurturing relationships with the targets to gain their trust (Aldawood & Skinner, 2020). The 'Trust' phase is crucial in establishing a relationship of trust between the attacker and the victim (Gallegos-Segovia et al., 2017). This phase exploits the human tendency to share information with trusted sources, allowing for the acquisition of critical data (Gallegos-Segovia et al., 2017; Aldawood & Skinner, 2020). The attacker initiates this phase, also known as the 'hook,' through direct contact or email communication (Salahdine & Kaabouch, 2019). The 'Exploitation' stage encompasses persuasion and manipulation, where the attacker gains access to sensitive information or coerces the target into committing security errors (Aldawood & Skinner, 2020). Upon establishing trust, the attacker exploits this connection by eliciting information or engaging the victim in specific actions. Emotional manipulation is often utilized to influence the victim into sharing sensitive data or committing security oversights (Gallegos-Segovia et al., 2017; Salahdine & Kaabouch, 2019). The last phase involves implementing the attack to obtain necessary information while attempting to erase any evidence that could link back to the attacker (Gallegos-Segovia et al., 2017; Aldawood & Skinner, 2020).

The Social Engineering model, notably defined by Kevin Mitnick in "The Art of Deception: Controlling the Human Element of Security" (Mitnick & Simon, 2002), comprises four core phases: research, establishment of rapport and trust, exploitation of trust, and utilization of information. This framework has been further elaborated in the work by Mouton et al. (2016), providing a comprehensive expansion on each phase of the social engineering attack. According to their ontological model, a social

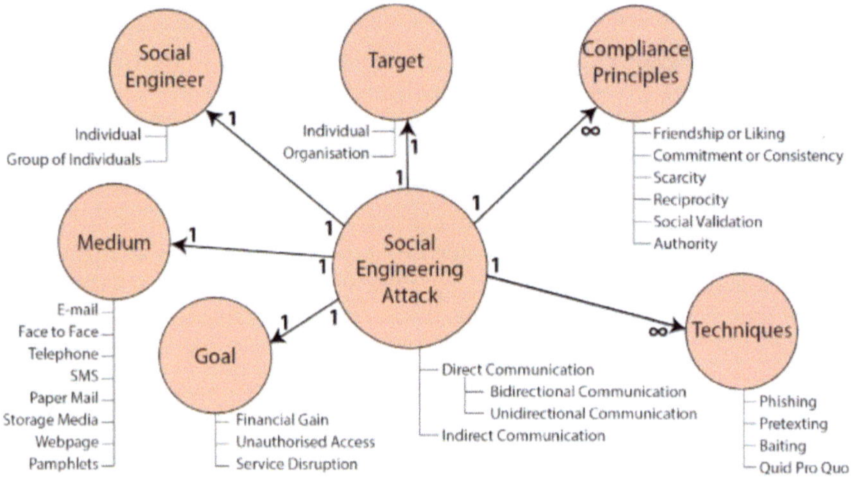

Fig. 1 Ontological model of a social engineering attack (Mouton et al., 2016)

engineering attack involves direct or indirect communication and comprises several key elements: a social engineer, a target, a medium, a defined goal, one or more compliance principles, and one or more techniques (Mouton et al., 2016). Additionally, the attack can be segmented into multiple phases, each phase treated as a distinct attack within the model's framework (Fig. 1).

The ontological model encompasses various essential components: a goal, a medium, a social engineer, a target, compliance principles, and techniques. Goals within an attack might encompass financial gain, unauthorized access, or service disruption. The medium denotes the communication channel, including methods such as email, face-to-face interaction, or telephone calls. Social engineers could be individuals or groups. Targets encompass both individuals and organizations. Compliance principles represent the reasons behind a target's compliance with the attacker's requests, while techniques refer to the strategies employed in social engineering attacks (Mouton et al., 2016). In the context of social engineering as a method, Gallegos-Segovia et al. (2017) describe four specific attack types that can be outlined from the phases proposed by Mitnick (Mitnick & Simon, 2002).

- Technical Attacks: These involve no direct interaction but use deceptive tools like false emails or infected files, impersonating reputable sources to extract information.
- Ego Attacks: Here, the attacker directly engages with the victim, showcasing intelligence to subtly manipulate and gather information.
- Sympathy Attacks: These attacks build trust through communication techniques, leading victims to share information based on trust and empathy.
- Bullying Attacks: Employing intimidation and coercion, this method instils fear to extract information effectively.

This section elucidates the persistent susceptibility of individuals to manipulative incursions, persisting despite the advancements in technological fortification and security protocols. This ongoing vulnerability often culminates in breaches and the exfiltration of data, demonstrating a palpable financial impact evidenced by the escalating costs associated with data breaches. The delineated procedural phases of social engineering attacks, whether human-centric or technology-based, underscore the pivotal role of trust and manipulation. Subsequent progression will delve into an exploration of the convergence between social engineering attacks and the realm of deepfake technology, elucidating how the deployment of deepfakes amplifies human susceptibility, thereby augmenting the efficacy of these attacks. This inquisition will illuminate the evolving landscape of cybersecurity threats, advocating the necessity for robust and adaptive countermeasures.

2.5 *Deepfakes in Social Engineering*

- Impact on Cybersecurity

The integration of deepfakes in Social Engineering Attacks (SEAs) profoundly disrupts not only cybersecurity but also elevates the difficulty in identifying manipulated content. The adoption of deepfakes within SEAs complicates cybersecurity efforts by amplifying the difficulty in identifying fabricated content. These artificially generated pieces of media, be it video or audio, have advanced to a level where they seamlessly replicate human appearances and voices. The nuances of facial expressions, speech patterns, and contextual authenticity in these fabricated contents often appear to mirror reality, challenging traditional detection methods (Albahar & Almalki, 2019; van der Sloot & Wagensveld, 2022). These manipulations craft personas or scenarios so realistic that they effectively deceive individuals, presenting video calls or conversations that are presumed from trustworthy sources or are authentic, coercing them into divulging sensitive information or performing unusual actions (Mirsky & Lee, 2021; Mitnick & Simon, 2002; Stupp, 2019). Incorporating deepfakes into traditional social engineering strategies fortifies the multifaceted nature of these attacks, exploiting emotional, psychological, and behavioural triggers to deceive individuals more effectively (BBC News, 2012), increasing success rates in acquiring sensitive information or manipulating targets for malicious ends.

- Erosion of Trust and Credibility

Holiday (2018) in his book vividly illustrates how vulnerable individuals are to media manipulation, especially in a landscape flooded with online misinformation and sensationalized headlines, exacerbating the challenge of preserving trust and credibility in these digital interactions. The nuanced fabrications facilitated by deepfakes foster a misleading sense of authenticity in these interactions, blurring the lines between what is genuine and what is fabricated. As these deceptive communications become increasingly sophisticated, they erode the foundations of trust and credibility that digital interactions rely upon (Holiday, 2018; Vaccari & Chadwick, 2020). The ability of these manipulations to replicate real-life scenarios with such accuracy raises significant concerns. It not only influences the ease with which individuals are deceived but also complicates the task of distinguishing between what is genuine and what is manipulated. This erosion of trust amplifies the complexities of countering such deceptive practices, especially considering the efficiency of these fabrications in imitating trusted or authoritative sources (Schwartz, 2018). The axiom 'seeing is believing' no longer stands, as distinguishing deepfakes from authentic content becomes increasingly challenging, albeit certain discernible features persist due to the foundational principles of AI algorithms. The evolution and refinement of these algorithms and datasets progressively augment the realism of deepfakes, significantly simplifying and enhancing the credibility of Social Engineering Attacks.

2.6 Real-World Scenarios

The seriousness of the Deepfake threat gained prominence with DARPA's recognition that even individuals lacking technical expertise could manipulate visual media (Siekierski, 2019). This evolution notably shifted social engineering tactics, allowing fraudsters to assume the identities of familiar figures, exploiting trust, and sidestepping usual scepticism in their interactions, thus posing as friends or colleagues without raising suspicion regarding their intentions. Instances in 2019 and early 2020 underscore the potency of these tactics, where voice cloning facilitated a successful deception of high-value transfers by impersonating executives (Brewster, 2021; Stupp, 2019). Notably, the case of "Vickie," a deepfake profile on LinkedIn, demonstrates the potential for these fabricated personas to engage with numerous sensitive individuals, raising concerns about diverse malicious purposes, including social engineering attacks, phishing scams, and espionage (Tomlinson, 2021). The advancement in deepfake quality reduces the reliance on physical alterations like makeup or costumes, exemplified in cases where individuals defrauded substantial sums by impersonating well-known public figures like the French Foreign Minister (Willsher & Holmes, 2019). It was arguably a Belgian political party's intentionally low-quality deep fake video of President Donald Trump appearing to tell Belgians that they should withdraw from the Paris climate change agreement, posted to social media in May 2018, that demonstrated how easily the technology could be used to mislead voters (Schwartz, 2018). It provoked hundreds of comments, many expressing outrage that the American president would dare weigh in on Belgium's climate policy. These vulnerabilities extend to instances like mass spear-phishing campaigns powered by deepfakes, where AI is utilized to clone voices and exploit trust for malicious intents. Malicious cyber actors may use synthetic content to create highly believable spear phishing messages or engage in sophisticated social engineering attacks, according to EUROPOL (2021). Similarly, advancements in AI technology have led to the creation of "Master Prints," deepfake fingerprints capable of bypassing partial print scanners, and adversarial samples that can deceive face and voice recognition systems further underline the severe challenges posed by the integration of deepfakes in SEAs.

2.7 *Current State*

FBI (2021) warns that Malicious Actors Almost Certainly Will Leverage Synthetic Content for Cyber and Foreign Influence Operations which can have a more severe and widespread impact due to the sophistication level of the synthetic media used. The utilization of Deepfakes in malicious activities remains limited to specialized groups or state-backed entities, with documented successful cases being rare, as noted by Hill (2021). Hill (2021) suggests an inevitable expansion in their usage, much like other technologies, once the tools, methods, and potential benefits become more widely recognized and disseminated. Mirsky and Lee (2021) support this by anticipating an upsurge in deepfake phishing attacks and scams, signifying a trend of weaponization of deepfakes for monetization purposes, with advancing tools becoming more practical and efficient in their deployment for both individuals and companies.

2.8 *Research Gap*

The current body of research presents notable gaps regarding the convergence of deepfake technology and social engineering attacks within the realm of cybersecurity. While extensive studies exist on deepfakes (Borji, 2023; Groh et al., 2021; Liu et al., 2023; Mirsky & Lee, 2021; Seow et al., 2022) as well as social engineering attacks (Aldawood & Skinner, 2020; Gallegos-Segovia et al., 2017; Koyun & Al Janabi, 2017; Krombholz et al., 2015; Mouton et al., 2016), limited research delves into the intricate interplay and impact of deepfakes on the efficacy of social engineering attacks. Current deepfake research can be categorized into two primary domains:

- Deepfake Generation: This area centres on enhancing output resolution while minimizing dataset, computational, and training resource requirements.
- Deepfake Detection: Research in this field prioritizes the development of resilient and versatile detectors for real-time scenarios.

The currently existing gaps result from the fast expansion of these technologies, which has outpaced scholarly investigations. Deepfake-generating techniques' increasing sophistication and incorporation into social engineering attacks have surpassed comprehensive scholarly research. The rapid advancement of these technologies poses new hurdles in recognising and fighting these composite threats. Furthermore, the multidisciplinary nature of this field adds to the gaps. Understanding deepfakes necessitates knowledge of artificial intelligence, image processing, and data science, but additionally understanding social engineering attacks requires knowledge of psychology, human behaviour, and cybersecurity. Bridging these disparate domains is difficult, resulting in fragmented research without an extensive, holistic viewpoint. Additional research is necessary in order to fully comprehend the scope and consequences of these interrelated dangers. Disclosing the complex relationships between social engineering and deepfakes will strengthen mitigation, prevention, and detection tactics. Investigating these gaps will improve detection systems and contribute to our knowledge of the psychological foundations that make people vulnerable to such deceptive techniques.

The selected research questions are directly related to these gaps in existing research. They seek to bridge the gap by investigating how deepfake technologies enhance the success rates of social engineering attacks, discovering vulnerabilities, and devising complete counterstrategies to these increasingly complex cyber threats. The study hypothesis posits that including deepfakes in social engineering attacks greatly increases the success rates of these deceptive tactics, emphasising the importance of adaptive responses in the expanding cybersecurity landscape.

3 Deepfake Creation

To elucidate the intricate process of crafting deepfakes through DeepFaceLab, the following systematic steps were rigorously followed:

- Download of DeepFaceLab:

Obtained the DeepFaceLab software from the official website, opting for the free version to facilitate experimentation. The link used is 'https://github.com/iperov/DeepFaceLab' (Fig. 2).

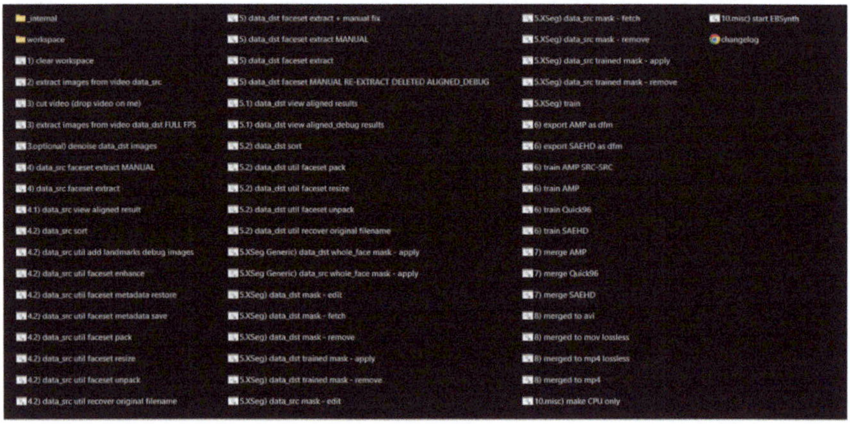

Fig. 2 DeepfaceLab downloaded files

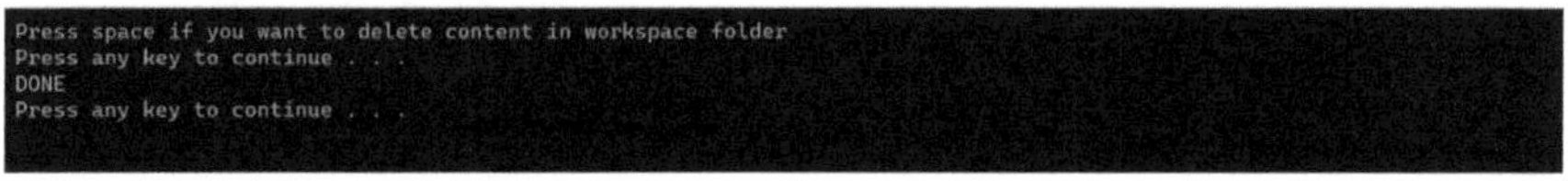

Fig. 3 Clearing the workspace

- Clear Workspace Initialization:

Executed the clear workspace batch file to ensure a pristine starting point, eliminating any residual artefacts from prior attempts (Fig. 3).

- Source Video Data Extraction:

Implemented the 'extract images from video data_src' batch file, initiating the extraction of frames from the source video. The source video is the video from which the face will be extracted and merged with the destination video, to create a deepfake. Default settings were employed for this step and all the following steps, highlighting the accessibility of deepfake creation even for those with limited technical expertise (Figs. 4, 5).

- Destination Video Data Extraction:

Utilized the 'extract images from video data_dst FULL FPS' batch file to extract frames from the destination video, storing the extracted faces conveniently in the workspace under the data_dst folder (Figs. 6, 7).

- Faceset Extraction from Source Video Images:

Executed the 'data_src faceset extract' command, extracting the faceset from images derived from the source video. This step contributes to a comprehensive understanding of the deepfake creation process (Figs. 8, 9).

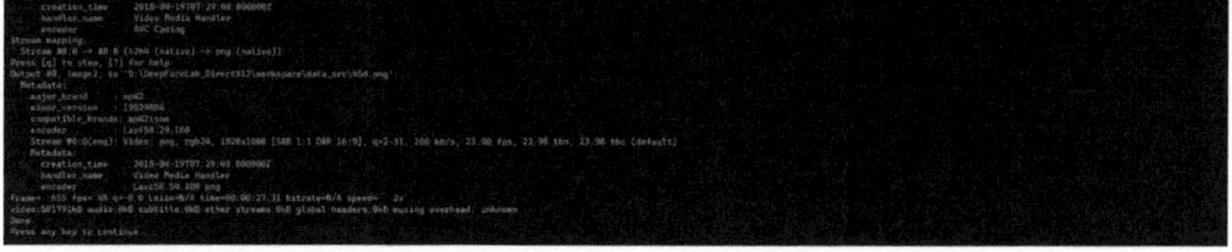

Fig. 4 Extracting frames from the source video

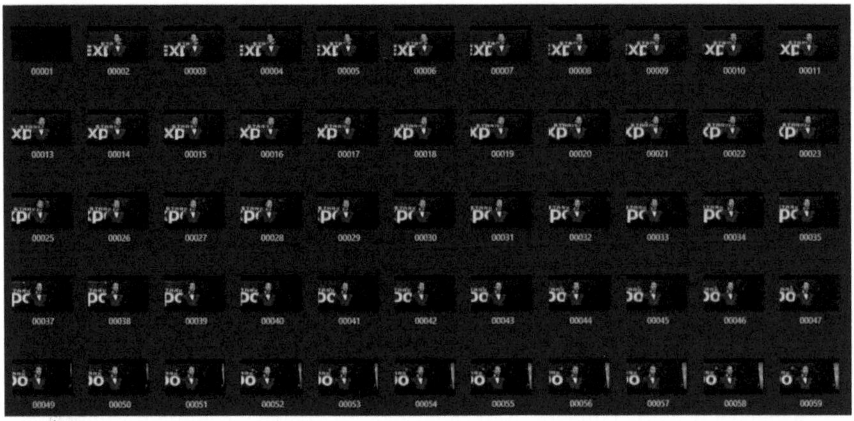

Fig. 5 Extracted frames

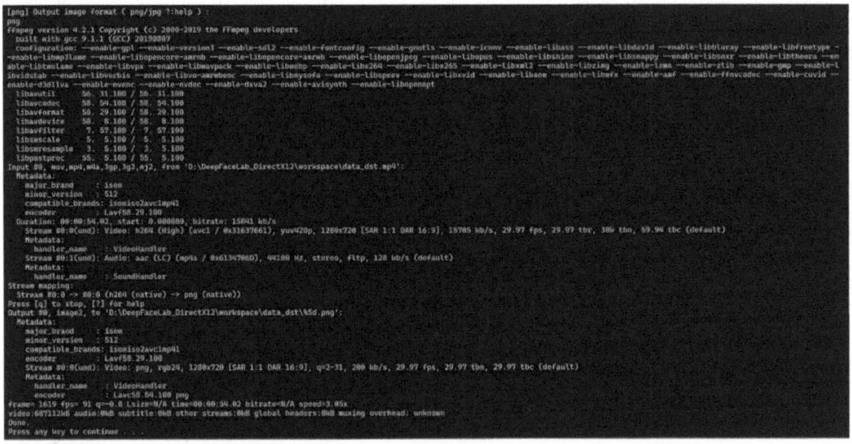

Fig. 6 Extracting frames from the destination video

- Destination Video Faceset Extraction:

Ran the 'data_dst faceset extract' batch file to extract the faceset from images derived from the destination video, storing both source and destination facesets under the 'aligned' folder for subsequent processing (Figs. 10, 11).

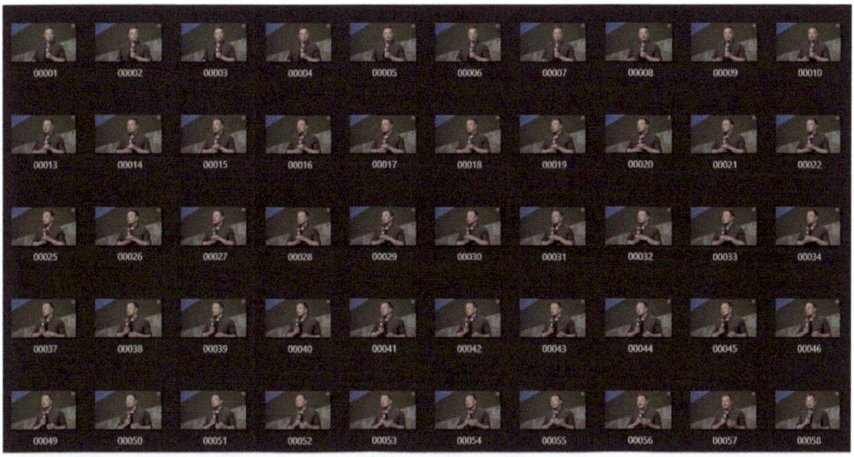

Fig. 7 Extracted frames

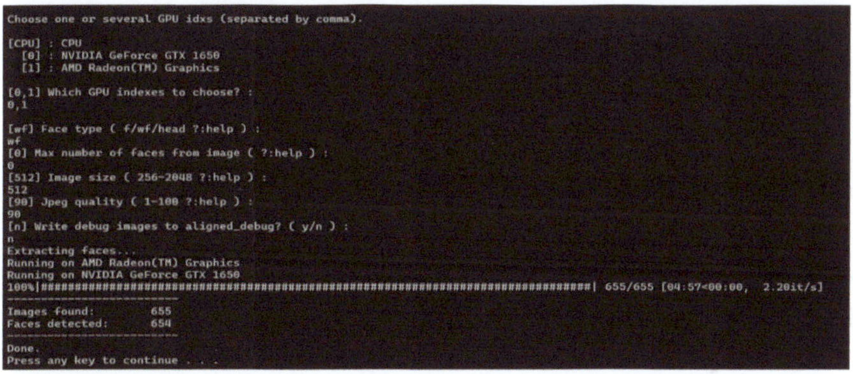

Fig. 8 Extracting face sets from the source frames

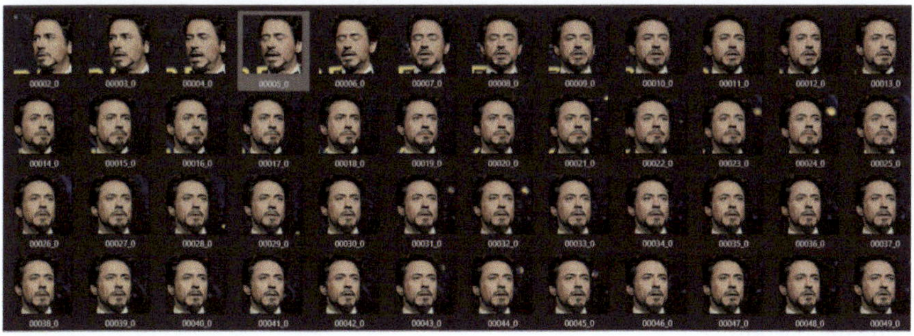

Fig. 9 Extracted facesets

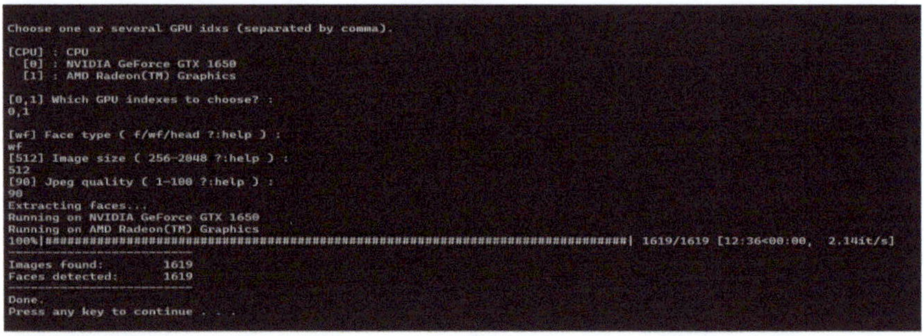

Fig. 10 Destination faceset extraction

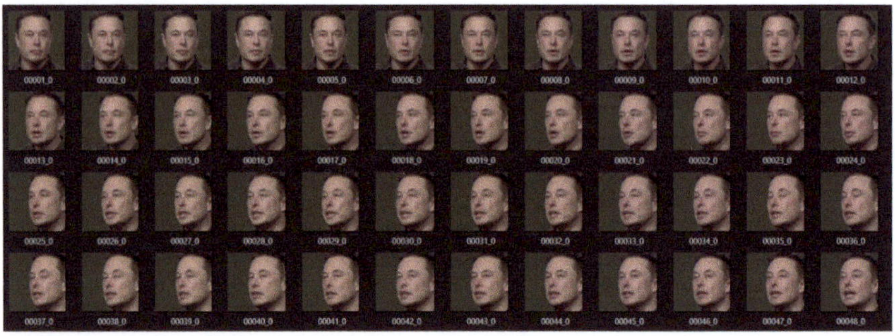

Fig. 11 Extracted facesets

3.1 *Model Training with 'Quick96'*

Initiated model training using the 'train Quick96' batch file. The term 'Quick96' denotes a specific model configuration, balancing speed, and quality. Multiple iterations were conducted, monitoring the training graph for convergence (closer to zero indicating better performance) (Figs. 12, 13).

- Faceset Integration using 'Quick96':

Executed the 'merge Quick96' batch file to integrate facesets using the trained model, contributing to the synthesis of a cohesive and realistic deepfake outcome (Figs. 14, 15, 16).

Two specific parameters, namely the erode_mask_modifier set to 20 and blur_mask_modifier set to 100, are adjusted. This deliberate modification serves the following purposes:

The increase in the erode_mask_modifier value to 20 intensifies the erosion applied to the mask during integration. Erosion involves the shrinking or removal of pixel clusters from the mask's boundaries. This adjustment can enhance the blending

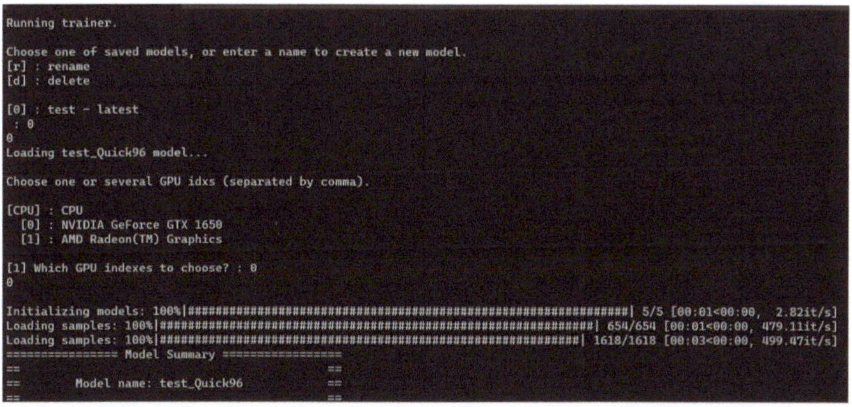

Fig. 12 Training the model

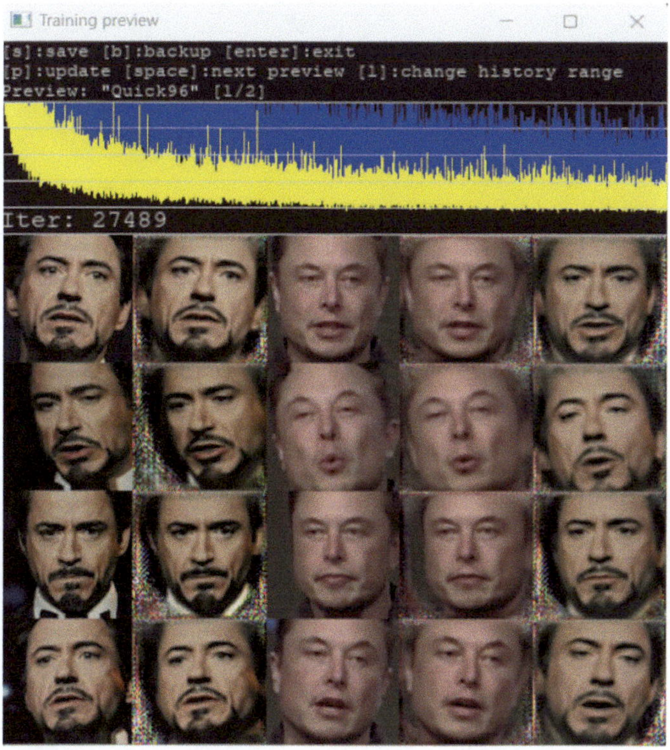

Fig. 13 Training preview

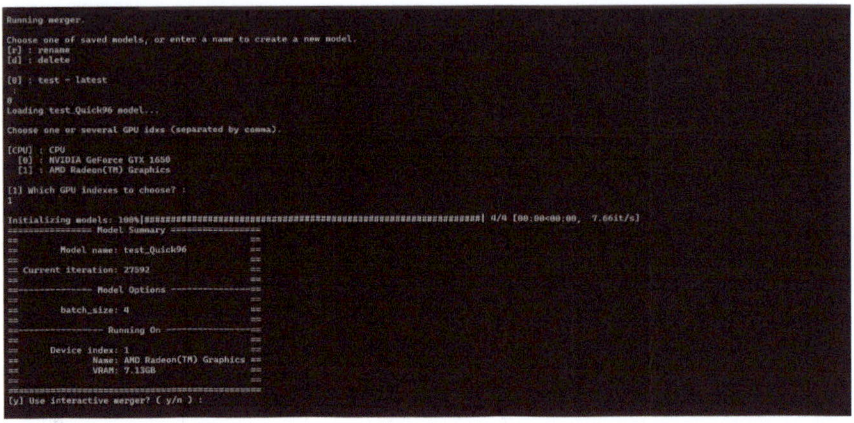

Fig. 14 Merger

Fig. 15 Default merger values

Fig. 16 Merger preview

of the face set with the destination video, particularly by refining the alignment of facial features and contours.

Setting the blur_mask_modifier to 100 significantly amplifies the blurring effect applied to the mask. This heightened blurring serves to mitigate any residual artifacts or sharp transitions in the integrated faceset, fostering a smoother and more natural appearance in the final deepfake output (Fig. 17).

- Conversion to MP4 Format:

Utilized the 'merged to mp4' batch file to convert the integrated result into an MP4 file. The final deepfake video, featuring merged faces from the source and destination, was stored in the workspace as 'result.mp4' (Fig. 18).

These steps collectively illustrate the comprehensive journey from initial extraction to model training, faceset integration, and the ultimate creation of a deepfake video in a user-friendly and accessible manner, even for those with limited technical proficiency. The iterative model training process ensures the continual enhancement of deepfake realism.

Fig. 17 Setting the mask values

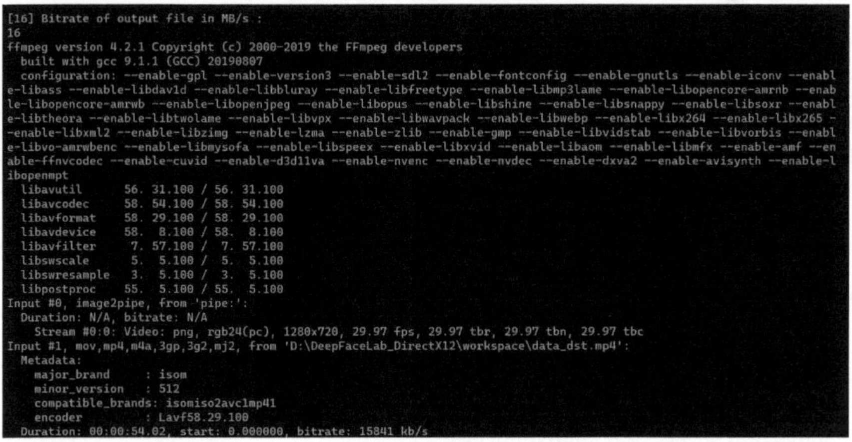

Fig. 18 Conversion to MP4

4 Deepfake Creation Process Insights

The process of creating deepfakes using DeepFace Lab involves crucial steps related to data extraction and faceset management, providing insightful considerations for these aspects:

- Data Selection and Preprocessing:

The choice of source material significantly influences the quality and realism of the deepfake. Selecting high-resolution, well-lit videos enhances the final output.

- Face Extraction and Alignment:

Precision in Face Selection: Accurate extraction of facial features is pivotal. DeepFace Lab employs facial landmark detection to precisely identify and extract key facial components from source images or videos.

- Consistent Alignment:

Proper alignment ensures coherence in facial movements across frames, minimizing distortions and improving the overall quality of the deepfake.

- Faceset Creation and Organization:

Diversity in Facesets: Utilize a diverse set of images for the target face to enhance adaptability to various expressions and lighting conditions, contributing to a more realistic deepfake.

Organized Dataset: Systematically structuring the faceset facilitates efficient training. Categorize images based on expressions, poses, or lighting conditions to enhance the model's learning.

Varied Expressions and Poses: Augment the dataset with images featuring different facial expressions, poses, and lighting scenarios. This broadens the model's understanding, enabling it to handle a wider range of inputs.

- Quality Control and Iterative Refinement:

Continuous Evaluation: Periodically assess the quality of the faceset and adjust parameters as needed. Iteratively refine the dataset by removing outliers or low-quality images to enhance the model's performance.

Equal Representation: Strive for an equal representation of both source and target faces to prevent biases in the deepfake output. Balancing the dataset contributes to more accurate and unbiased results.

- Privacy and Security Measures:

Face Anonymization: Prioritize the anonymization of faces not intended for manipulation to uphold privacy standards.

Secure Data Storage: Implement secure storage practices for the dataset to prevent unauthorized access or potential misuse.

Insights into data extraction and faceset management underscore the importance of meticulous preparation and ethical considerations throughout the deepfake creation process. These considerations not only impact the quality of the final output but also play a crucial role in addressing ethical concerns associated with the use of synthetic media.

- Enhanced Technical Proficiency:

The immersive engagement with DeepFaceLab has significantly bolstered technical proficiency in the creation of deepfakes. This hands-on experience not only helps the author understand the complexities but also provides a comprehensive understanding of the intricacies inherent in generating manipulated media content.

- User-Friendly Interface Promoting Accessibility:

The recognition of DeepFace Lab's user-friendly interface emphasizes its potential to democratize deepfake creation, extending accessibility to individuals with diverse backgrounds and limited technical knowledge. The numbered batch files containing the code facilitate usage without the need for users to comprehend the intricate backend processes. However, this accessibility to potentially hazardous software raises concerns, especially if the tool falls into the hands of inexperienced users, such as children. Drawing from personal childhood experiences, where edited pictures were employed for bullying, the transition to deepfakes intensifies the risk due to their heightened realism and simplified creation process. The combination of increased realism and ease of use poses challenges in mitigating the potential misuse of deepfake technology, necessitating a closer examination of ethical considerations and protective measures to address the associated societal risks.

- Model Training and Optimization:

The exploration of model training, particularly with the 'Quick96' configuration, has elucidated the delicate balance between speed and quality in the deepfake creation process. This profound insight into model optimization serves as a cornerstone for achieving heightened realism in the generated deepfakes.

- Holistic Experience in Faceset Integration and Video Conversion:

The practical experience of seamlessly integrating facesets using the trained model and converting the final result into an MP4 format provides a holistic view of the entire deepfake creation journey. This hands-on familiarity enriches the understanding of the intricate technical aspects involved.

- Navigating Challenges in Parameter Fine-Tuning:

The encountered challenges in fine-tuning the parameters during the deepfake creation process highlight the critical role of precise parameter selection in achieving realistic outcomes. Addressing these challenges not only hones the technical aspects of the creation process but also contributes to the continuous enhancement of deepfake creation techniques. The quality of the generated result could have been enhanced by filtering undesirable frames such as the frames that lack a discernible face or featuring a blurry countenance.

This experiential learning has emphasized the significance of iterative model training for elevating deepfake quality. This iterative approach allows for continual adjustments and enhancements, highlighting the dynamic nature of deepfake creation and the continual pursuit of enhanced realism. Extending the model's iterations could further enhance the realism of the deepfake.

Another insight gained from this creation process reveals considerations regarding the similarity of skin colour and body shape. As DeepFace Lab focuses solely on facial replacement, if the face of the source video is either lighter or darker in colour than the destination video, it may result in incongruities when the source video's face differs in colour from the destination video or when there is a mismatch in body characteristics, leading to perceptible dysmorphia in the composite output.

The profound insights gained from the practical experience intricately inform the ongoing research, providing a nuanced understanding of the technical dimensions of deepfake creation. This practical knowledge serves as a robust foundation for the meticulous analysis and assessment of the profound impact of deepfake-driven social engineering attacks on cybersecurity.

5 Deepfakes in Social Engineering Attacks

- Scenario 1—IT helpdesk

Within the realm of Social Engineering Attacks (SEA), the scenario proposed by Flick and Morehouse (2011) involves impersonating an employee to manipulate their password through the IT Help Desk, presenting a tangible threat heightened by the advent of deepfake technology. This sophisticated technique exploits vulnerabilities within human interactions and organizational structures, emphasizing the pressing need for cybersecurity measures.

In this scenario, malicious actors could employ deepfake technology to convincingly replicate the voice and appearance of a legitimate employee seeking assistance from the IT Help Desk. Deepface Lab, a tool explored in prior research, enables the creation of highly realistic deepfakes by leveraging deep learning algorithms. Through the synthesis of authentic-looking facial expressions, voice modulations, and even subtle behavioural nuances, the deepfake could successfully imitate the targeted employee, making it challenging for IT personnel to discern the deception.

oracle mind (2016) demonstrates the exploitation of psychological factors, exemplified by the use of the sound of a baby crying and a distressed woman impersonating the employee's wife to illicitly obtain a password reset link from the IT help desk. So even without the actual person being the one to make the call, the attacker was able to get the login details. But with the use of deepfake, the attacker would not need to go to such lengths to gain the login details.

The attacker initiates the exploit by contacting the IT Help Desk, posing as the impersonated employee facing login issues or claiming to have forgotten their password. Leveraging the deepfake, the imposter adeptly mimics the targeted individual's speech patterns, tone, and overall demeanour, adding a layer of authenticity to the ruse. The fabricated urgency of the situation might pressure the IT personnel to expedite the process, bypassing standard verification protocols.

Given the apparent legitimacy of the impersonation, the IT Help Desk may unwittingly assist the attacker in resetting the password or granting unauthorized access to sensitive systems. This manipulation could lead to unauthorized data access, corporate espionage, or the compromise of critical information, posing severe repercussions for the targeted organization.

The integration of deepfake technology into this SEA scenario elevates the threat level significantly. The ability to replicate both visual and auditory elements with remarkable accuracy enhances the effectiveness of the social engineering attack. As organizations continue to fortify their defences against conventional impersonation tactics, the integration of deepfakes introduces a new layer of complexity that demands heightened awareness, robust authentication procedures, and employee training to recognize and mitigate these evolving cyber threats.

Successful deepfake-enabled attacks on IT security pose severe consequences, including the risk of unauthorized access and potential data compromise. The integration of deepfake technology into Social Engineering Attacks (SEA) amplifies the sophistication of impersonation tactics, making it challenging for IT personnel to discern deception.

- Countermeasures and best practices for IT personnel:

  The following are some good countermeasures and best practises:

- Multi-Factor Authentication (MFA): Implement robust MFA systems to add an additional layer of verification, reducing the likelihood of unauthorized access even in the presence of convincing deepfakes.
- Behavioural Analysis: Leverage behavioural analysis tools to detect anomalies in user behaviour, helping identify discrepancies between genuine employees and potential deepfake impersonators.
- Employee Training Programs: Conduct regular training sessions to educate IT personnel on recognizing signs of deepfake manipulation. This includes emphasizing the importance of verifying requests for sensitive information, even when they appear to come from familiar sources.
- Enhanced Verification Protocols: Establish stringent verification protocols for high-risk actions, such as password resets or access to sensitive information. Encourage IT staff to follow these protocols diligently, regardless of the urgency conveyed by the requester.
- Real-time Communication Channels: Implement secure and verified communication channels for handling sensitive requests. This ensures that legitimate employees can authenticate their identity through channels that are less susceptible to deepfake exploitation.

By adopting the suggested countermeasures, organizations can fortify their defences against deepfake-enabled impersonation attempts, safeguarding IT security and mitigating the potential risks associated with such attacks.

- Scenario 2—Political

In the scenario of political manipulation, the nefarious potential of deepfakes casts a considerable shadow, introducing a malicious dimension to information warfare. Holiday (2018) elucidates how vulnerable individuals are to media manipulation. Misinformation is already prevalent in electoral campaigns (Grossman & Helpman, 2022). Hochschild and Einstein (2015) throw light on how the electors consider the misinformation disseminated during the electoral campaigns when deciding which candidate to vote for. The electors choose to believe the rumours spread without bothering to verify if they are factual. Now consider a scenario wherein a meticulously crafted deepfake video surfaces, portraying a prominent political figure engaging in condemnable acts or delivering incendiary statements. The fabricated content, appearing authentic in its presentation, disseminates rapidly across social media platforms, instigating widespread public outrage and amplifying divisive sentiments.

In this hypothetical situation, the malicious entities orchestrating the deepfake strategically unleash the video amid a pivotal election campaign. The objective is evident: to destabilize the political landscape, sow discord among voters, and undermine the credibility of the targeted political figure. The carefully designed video, exploiting existing societal tensions, possesses the potential to sway public opinion, besmirch reputations, and ultimately impact the electoral outcome. Parkin (2019) illustrates a case of deepfake utilization in politics by presenting an example where Trump seems to mock Belgium for maintaining its commitment to the Paris climate agreement. Given the predisposition of the public to accept hearsay without factual verification, the advent of deepfakes exacerbates this tendency, leading individuals to perceive manipulated content as authentic. The repercussions transcend the immediate political domain, permeating public trust in institutions and the democratic process. As the deepfake gains momentum, traditional media outlets encounter challenges in real-time verification of content authenticity, exacerbating the difficulty of countering the dissemination of misinformation. Political adversaries seize the opportunity to capitalize on the ensuing chaos, further escalating polarization within the electorate.

- Implications

Deepfake-driven misinformation in political contexts has the following implications:

- Trust Erosion: Deepfake-driven misinformation contributes to a decline in public trust, as individuals may struggle to differentiate between authentic and manipulated content.
- Political Polarization: The strategic use of deepfakes exacerbates political polarization, fostering an environment where misinformation fuels division and hampers constructive political discourse.

- Crisis of Confidence: Public confidence in institutions and democratic processes may be shaken, as manipulated content challenges the integrity of political figures and electoral systems.

This scenario underscores the imperative to comprehend and address the multifaceted threats posed by deepfakes in political contexts. It accentuates the necessity for robust detection mechanisms, legislative measures, and public awareness campaigns to alleviate the potential harm inflicted by manipulated media. Edelman (2023) discusses lagging in efforts to address political deepfakes ahead of the 2024 election and how only three states in the US passed laws in 2023 to address deepfakes. Stakeholders are urged to strengthen defences against the weaponization of synthetic media in the quest for political influence, as this situation serves as a harsh reminder of the delicate balance between technological progress and the preservation of democratic processes. Furthermore, it underscores the importance of cultivating digital literacy to empower individuals in critically assessing the veracity of online content, particularly in the turbulent terrain of contemporary politics.

6 Ethical Considerations

Deepfakes, with their ability to manipulate visual and auditory content seamlessly, raise profound ethical concerns that demand rigorous examination. In the realm of social engineering attacks, the ethical dimensions of impersonation and deception come to the forefront. The creation of deepfakes for malicious purposes, such as mimicking an employee's identity to exploit IT Help Desk services, challenges the boundaries of trust within organizational structures. The unauthorized use of deepfakes in this context undermines the integrity of communication channels, potentially leading to unauthorized access and data breaches. Ethical considerations extend to the potential psychological impact on individuals whose identities are exploited, highlighting the need for safeguards to protect personal and professional reputations.

In the political landscape, the ethical implications of deepfake use manifest in the weaponization of synthetic media to manipulate public opinion. Fabricated videos portraying political figures engaging in reprehensible acts pose a substantial threat to the democratic process. The dissemination of misleading information through deepfakes jeopardizes the foundational principles of informed decision-making in elections. Addressing these ethical concerns involves recognizing the potential for widespread societal distrust and polarization resulting from manipulated content. Responsible deepfake practices necessitate transparency in media dissemination, rigorous fact-checking processes, and public awareness campaigns to foster digital literacy.

Societal risks associated with unethical deepfake practices include the erosion of trust in digital media, increased susceptibility to misinformation, and potential harm to individuals and institutions. Emphasizing responsible deepfake practices involves advocating for robust authentication mechanisms, stringent regulations, and the integration of ethical considerations into deepfake detection technologies. Striking a balance between technological advancements and ethical responsibilities is imperative to prevent the malicious misuse of deepfakes and to foster a digital landscape grounded in trust, integrity, and informed decision-making. As technology evolves, a proactive ethical framework becomes indispensable in mitigating the ethical challenges posed by deepfakes.

7 Countermeasures, Gaps and Solutions

- Deepfake Countermeasures

In the realm of cybersecurity, current measures employ technical detection, fact-checking initiatives, and media literacy education. Technical detection, leveraging machine learning algorithms, showcases strengths in identifying inconsistencies in deepfake content, such as lip-syncing and skin tone abnormalities. However, its limitations include reduced accuracy against sophisticated models, susceptibility to adversarial attacks, and high computational costs. Fact-checking initiatives, marked by rigorous investigation and analysis, contribute to exposing falsehoods, building public awareness, and fostering trust in reliable sources. Despite these strengths, they face challenges related to limited reach, susceptibility to fatigue and disinformation campaigns, and the difficulty of verifying rapidly evolving content. Media literacy education stands as a long-term strategy, empowering individuals to critically evaluate information and build resilience against misinformation. While it holds promise for societal impact, its weaknesses lie in the slow diffusion of knowledge, difficulty reaching all audience segments, and insufficient resources for comprehensive education programs.

On the political front, measures include deepfake regulation, transparency, and accountability efforts, and international cooperation. Deepfake regulation carries the potential to deter malicious actors, promote transparency, and protect public discourse. However, it grapples with concerns related to free speech, potential censorship, and the challenge of defining and enforcing regulations across diverse contexts. Transparency and accountability initiatives contribute to building trust in institutions and combating public scepticism. Yet, their effectiveness depends on a strong commitment from all actors and is vulnerable to manipulation and selective disclosure, risking potential backlash from vested interests.

International cooperation emerges as a strength in enabling knowledge sharing, facilitating coordinated responses to global threats, and promoting harmonization of laws and standards. However, difficulties arise in overcoming political and cultural differences, establishing frameworks for collaboration, and navigating bureaucratic slowness.

- Gaps and Potential Solutions:

Identified gaps encompass technology, information flow, legal and ethical frameworks, and global coordination. Solutions involve advanced detection technology, bolstering information ecosystem resilience, establishing a global task force, and legal harmonization.

Advanced detection tech solutions include developing context-aware AI algorithms, exploring blockchain-based authentication systems, and investing in open-source detection tools. For information ecosystem resilience, support for independent media initiatives, promotion of decentralized information sources, and encouragement of critical thinking skills in education are recommended.

8 Conclusions

In conclusion, this Research has delved into the intricate landscape of deepfake technology, exploring its applications in social engineering attacks on IT security and its potential ramifications in political manipulation. The research amalgamated theoretical insights from a comprehensive literature review with practical experiences gained through engaging with DeepFace Lab. Key findings and contributions can be summarized as follows:

Technological Understanding: Through a hands-on exploration of DeepFace Lab, a detailed understanding of the deepfake creation process was achieved. This practical knowledge served as a foundation for analysing the capabilities and challenges associated with synthetic media.

Scenario Development and Analysis: The formulation of hypothetical scenarios facilitated the examination of deepfake applications in diverse contexts. Scenario analyses underscored the multifaceted threats deepfakes pose in IT security and political discourse, emphasizing the importance of proactive countermeasures.

Ethical Considerations: Ethical dimensions of deepfake technology were scrutinized, particularly in terms of privacy, security, and potential misuse. The research highlighted the need for ethical guidelines and protective measures to mitigate societal risks.

Countermeasure Evaluation: Existing countermeasures against deepfake threats were evaluated, revealing gaps and areas for improvement. The assessment emphasized the dynamic nature of the deepfake landscape, necessitating ongoing research and adaptive defences.

Public Perception and Media Literacy: Investigation into public perceptions of deepfakes emphasized the significance of media literacy in mitigating the impact of synthetic media. Awareness and education were identified as pivotal components in addressing societal vulnerabilities.

Comparative Analysis: A comparative analysis of scenarios in IT security and political manipulation unveiled commonalities and differences, offering insights into the distinct challenges posed by deepfakes in varied domains.

Recommendations and Future Research: Actionable recommendations were provided for enhancing countermeasures and addressing identified vulnerabilities.

This work propose avenues for future research, emphasizing the need for continuous vigilance and collaboration in the face of evolving deepfake threats.

In an overview, this research adds to the existing debate around deepfake technology by giving a comprehensive evaluation of its technological complexities, ethical ramifications, and societal repercussions. The findings highlight the importance of multidisciplinary collaboration among researchers, politicians, and society to appropriately traverse the complicated landscape of artificially generated media. As technology progresses, a proactive and collaborative strategy is required to protect.

References

Aaron, Li, Y., & Vinyals, O. (2018). *Representation learning with contrastive predictive coding.* Cornell University. arXiv. https://doi.org/10.48550/arxiv.1807.03748

Albahar, M., & Almalki, J. (2019). Deepfakes: Threats and countermeasures systematic review. *Journal of Theoretical and Applied Information Technology, 97*, 22. [online]. https://www.jatit.org/volumes/Vol97No22/7Vol97No22.pdf

Aldawood, H., & Skinner, G. (2018). Educating and raising awareness on cyber security social engineering: A literature review. In *2018 IEEE International Conference on Teaching, Assessment, and Learning for Engineering (TALE).* [online]. https://doi.org/10.1109/tale.2018.8615162

Aldawood, H., & Skinner, G. (2020). An advanced taxonomy for social engineering attacks. *International Journal of Computer Applications, 177*(30), 1–11. https://doi.org/10.5120/ijca2020919744

BBC News. (2012, December 13). *Viewpoint: How hackers exploit 'the seven deadly sins'*. BBC News. [online]. https://www.bbc.co.uk/news/technology-20717773. Accessed 11 Nov 2023

Borji, A. (2023). Qualitative failures of image generation models and their application in detecting deepfakes. *Image and Vision Computing, 137*, 104771–104771. https://doi.org/10.1016/j.imavis.2023.104771

Brewster, T. (2021). Fraudsters cloned company director's voice in $35 million heist, police find. *Forbes.* [online]. http://forbes.com/sites/thomasbrewster/2021/10/14/huge-bank-fraud-uses-deep-fake-voice-tech-to-steal-millions/?sh=6fa0ce2d7559. Accessed 12 Nov 2023

Cassauwers, T. (2019). *Can artificial intelligence help end fake news? | Research and Innovation.* [online]. ec.europa.eu. https://ec.europa.eu/research-and-innovation/en/horizon-magazine/can-artificial-intelligence-help-end-fake-news

Chang, M. C., & Park, D. (2020). How can blockchain help people in the event of pandemics such as the COVID-19? *Journal of Medical Systems, 44*(5). https://doi.org/10.1007/s10916-020-01577-8

Chen, J., Mueller, J., Ioannidis, V. N., Goldstein, T., & Wipf, D. (2022). *A robust stacking framework for training deep graph models with multifaceted node features.* arXiv.org. [online]. https://doi.org/10.48550/arXiv.2206.08473

Chen, X. I., Mishra, N., Rohaninejad, M., & Abbeel, P. (2018). *PixelSNAIL: An improved autoregressive generative model.* [online]. proceedings.mlr.press. https://proceedings.mlr.press/v80/chen18h.html. Accessed 13 Jan 2024.

Chen, Y., Zhao, Y., Jia, W., Cao, L., & Liu, X. (2020). Adversarial-learning-based image-to-image transformation: A survey. *Neurocomputing, 411*, 468–486. https://doi.org/10.1016/j.neucom.2020.06.067

Chesney, R., & Citron, D. K. (2018). Deep fakes: A looming challenge for privacy, democracy, and national security. *SSRN Electronic Journal, 107*(1753). [online]. https://doi.org/10.2139/ssrn.3213954

Deepfakes. (2023). *deepfakes_faceswap*. GitHub. [online]. http://github.com/deepfakes/faceswap. Accessed 13 Nov 2023.

Department of Homeland Security. (2020). *DEEP increasing threat of AKE identities*. [online]. https://www.dhs.gov/sites/default/files/publications/increasing_threats_of_deepfake_identities_0.pdf. Accessed 6 Nov 2023.

Dong, C., Loy, C. C., He, K., & Tang, X. (2016). Image super-resolution using deep convolutional networks. *IEEE Transactions on Pattern Analysis and Machine Intelligence, 38*(2), 295–307. [online]. https://doi.org/10.1109/tpami.2015.2439281

Edelman, A. (2023). *States are lagging in tackling political deepfakes, leaving potential threats unchecked heading into 2024*. NBC News. [online]. https://www.nbcnews.com/politics/artificial-intelligence-deepfakes-2024-election-states-rcna129525

EUROPOL. (2021). *Malicious uses and abuses of Artificial Intelligence*. Europol. [online]. https://www.europol.europa.eu/publications-events/publications/malicious-uses-and-abuses-of-artificial-intelligence

FBI. (2021). *Malicious actors almost certainly will leverage synthetic content for cyber and foreign influence operations*. [online]. https://www.ic3.gov/Media/News/2021/210310-2.pdf. Accessed 12 Nov 2023

FBI. (2022). *Federal Bureau of Investigation: Internet crime report 2021*. FBI. [online]. https://www.ic3.gov/Media/PDF/AnnualReport/2021_IC3Report.pdf. Accessed 11 Nov 2023

Flick, T., & Morehouse, J. (2011). *Chapter 7—Attacking the utility companies*. ScienceDirect. [online]. https://www.sciencedirect.com/science/article/abs/pii/B9781597495707000078. Accessed 13 Nov 2023.

Galbally, J., Marcel, S., & Fierrez, J. (2014). Biometric antispoofing methods: A survey in face recognition. *IEEE Access, 2*, 1530–1552. [online]. https://doi.org/10.1109/ACCESS.2014.2381273.

Gallegos-Segovia, P. L., Bravo-Torres, J. F., Larios-Rosillo, V. M., Vintimilla-Tapia, P. E., Yuquilima-Albarado, I. F., & Jara-Saltos, J. D. (2017). *Social engineering as an attack vector for ransomware*. IEEE Xplore. [online]. https://doi.org/10.1109/CHILECON.2017.8229528.

Golić, Z. (2020). Finance and Artificial Intelligence: The Fifth Industrial Revolution and its impact on the financial sector. *ЗБОРНИК РАДОВА ЕКОНОМСКОГ ФАКУЛТЕТА У ИСТОЧНОМ САРАЈЕВУ, 8*(19), 67. [online]. https://doi.org/10.7251/zrefis1919067g

Grand View Research (2022). *Artificial Intelligence Market Size, Share | AI Industry Report, 2025*. Grand View Research. [online]. https://www.grandviewresearch.com/industry-analysis/artificial-intelligence-ai-market

Groh, M., Epstein, Z., Firestone, C., & Picard, R. (2021). Deepfake detection by human crowds, machines, and machine-informed crowds. *Proceedings of the National Academy of Sciences, 119*(1). https://doi.org/10.1073/pnas.2110013119

Grossman, G. M., & Helpman, E. (2022). Electoral competition with fake news. *European Journal of Political Economy, 77*, 102315. https://doi.org/10.1016/j.ejpoleco.2022.102315

Hill, M. (2021). *How deepfakes enhance social engineering and authentication threats, and what to do about it*. CSO Online. [online]. https://www.csoonline.com/article/571455/how-deepfakes-enhance-social-engineering-and-authentication-threats-and-what-to-do-about-it.html

Hochschild, J. L., & Einstein, K. L. (2015). Do facts matter? Information and misinformation in American politics. *Political Science Quarterly, 130*(4), 585–624. https://doi.org/10.1002/polq.12398

Holiday, R. (2018). *Trust me, i'm lying: Confessions of a media manipulator*. Profile Books.

Jones, M. (2013). Media-bodies and Photoshop. In *Controversial images* (pp. 19–35). https://doi.org/10.1057/9781137291998_2

Kanwal, N., Girdhar, A., Kaur, L., & Bhullar, J. (2019). Detection of digital image forgery using fast Fourier transform and local features. In *2019 International Conference on Automation, Computational and Technology Management (ICACTM)*. https://doi.org/10.1109/icactm.2019.8776709

Karras, T., Aila, T., Laine, S., & Lehtinen, J. (2017). *Progressive growing of GANs for improved quality, stability, and variation.* https://doi.org/10.48550/arxiv.1710.10196
Kietzmann, J., Lee, L. W., McCarthy, I. P., & Kietzmann, T. C. (2019). Deepfakes: Trick or treat? *Business Horizons, 63*(2), 135–146. [online]. https://doi.org/10.1016/j.bushor.2019.11.006
Kingma, D. P., & Welling, M. (2019). An introduction to variational autoencoders. *Foundations and Trends® in Machine Learning, 12*(4), 307–392. https://doi.org/10.1561/2200000056
Koyun, A., & Al Janabi, E. (2017). Social engineering attacks. *Journal of Multidisciplinary Engineering Science and Technology (JMEST), 4*(6), 2458–9403. [online]. https://www.jmest.org/wp-content/uploads/JMESTN42352270.pdf
Krombholz, K., Hobel, H., Huber, M., & Weippl, E. (2015). Advanced social engineering attacks. *Journal of Information Security and Applications, 22*, 113–122. [online]. https://doi.org/10.1016/j.jisa.2014.09.005
Li, P., Pei, Y., & Li, J. (2023). A comprehensive survey on design and application of autoencoder in deep learning. *Applied Soft Computing, 138*, 110176. https://doi.org/10.1016/j.asoc.2023.110176
Liu, H., Zhou, W., Chen, D., Fang, H., Bian, H., Liu, K., Zhang, W., & Yu, N. (2023). Coherent adversarial deepfake video generation. *Signal Processing, 203*, 108790. https://doi.org/10.1016/j.sigpro.2022.108790
Maddocks, S. (2020). 'A deepfake porn plot intended to silence me': Exploring continuities between pornographic and 'political' Deep Fakes. *Porn Studies, 7*(4), 1–9. https://doi.org/10.1080/23268743.2020.1757499
Microsoft. (2018). *Artificial Intelligence in Europe report: At a Glance.* Microsoft Pulse. [online]. https://pulse.microsoft.com/en/transform-en/na/fa1-articial-intelligence-report-at-a-glance/
Mirsky, Y., & Lee, W. (2021). The creation and detection of deepfakes. *ACM Computing Surveys, 54*(1), 1–41. https://doi.org/10.1145/3425780
Mitnick, K., & Simon, W. L. (2002). *The art of deception: Controlling the human element of security.* Wiley, Cop.
Mouton, F., Leenen, L., & Venter, H. S. (2016). Social engineering attack examples, templates and scenarios. *Computers & Security, 59*, 186–209. https://doi.org/10.1016/j.cose.2016.03.004
Nguyen, T. T., Nguyen, Q. V. H., Nguyen, D. T., Nguyen, D. T., Huynh-The, T., Nahavandi, S., Nguyen, T. T., Pham, Q.-V., & Nguyen, C. M. (2022). Deep learning for deepfakes creation and detection: A survey. *Computer Vision and Image Understanding, 223*(103525), 103525. [online]. https://doi.org/10.1016/j.cviu.2022.103525
Oord, A. V., Kalchbrenner, N., & Kavukcuoglu, K. (2016). *Pixel recurrent neural networks.* [online]. proceedings.mlr.press. https://proceedings.mlr.press/v48/oord16.html.
oracle mind. (2016). *This is how hackers hack you using simple social engineering.* YouTube. https://www.youtube.com/watch?v=lc7scxvKQOo.
Parkin, S. (2019). The rise of the deepfake and the threat to democracy. *The Guardian.* [online]. https://www.theguardian.com/technology/ng-interactive/2019/jun/22/the-rise-of-the-deepfake-and-the-threat-to-democracy.
Peltier, T. R. (2006). Social engineering: Concepts and solutions. *EDPACS, 33*(8), 1–13. https://doi.org/10.1201/1079.07366981/45802.33.8.20060201/91956.1
Rao, L., & Yang, J. (2020). Speech dereverberation based on improved wasserstein generative adversarial networks. *Journal of Physics: Conference Series, 1621*, 012089. https://doi.org/10.1088/1742-6596/1621/1/012089
Reuters Fact Check. (2023, February 1). Fact check-video of Biden's Ukraine speech was digitally altered. *Reuters.* [online]. http://reuters.com/article/factcheck-biden-stutter-idUSL1N34H1KL. Accessed 13 Nov 2023.
Rincon, L. (2023). *Virtually try on clothes with a new AI shopping feature.* Google. [online]. https://blog.google/products/shopping/ai-virtual-try-on-google-shopping/
Salahdine, F., & Kaabouch, N. (2019). Social engineering attacks: A survey. *Future Internet, 11*(4), 89. https://doi.org/10.3390/fi11040089

Schwartz, O. (2018). You thought fake news was bad? Deep fakes are where truth goes to die. *The Guardian.* [online]. https://www.theguardian.com/technology/2018/nov/12/deep-fakes-fake-news-truth

Seow, J. W., Lim, M. K., Phan, R. C. W., & Liu, J. K. (2022). A comprehensive overview of Deepfake: Generation, detection, datasets, and opportunities. *Neurocomputing, 513*, 351–371. [online]. https://doi.org/10.1016/j.neucom.2022.09.135

Siekierski, B. J. (2019). *Deep fakes: What Can be done about synthetic audio and video?* [online] lop.parl.ca. https://lop.parl.ca/sites/PublicWebsite/default/en_CA/ResearchPublications/201911E

Silbey, J., & Hartzog, W. (2019). Maryland law review the upside of deep fakes. *UMLL Journal, 78*(4). [online]. https://digitalcommons.law.umaryland.edu/cgi/viewcontent.cgi?article=3837&context=mlr

Statista (2023). *Cost of a data breach in the U.S. 2022.* Statista. [online]. https://www.statista.com/statistics/273575/us-average-cost-incurred-by-a-data-breach/

Stupp, C. (2019, August 30). Fraudsters used AI to Mimic CEO's voice in unusual cybercrime case. *Wall Street Journal.* [online]. http://wsj.com/articles/fraudsters-use-ai-to-mimic-ceos-voice-in-unusual-cybercrime-case-11567157402. Accessed 11 Nov 2023.

Tewari, A., Elgharib, M., Bharaj, G., Bernard, F., Seidel, H.-P., Perez, P., Zollhoefer, M., & Theobalt, C. (2020). *StyleRig: Rigging StyleGAN for 3D control over portrait images.* [online]. vcai.mpi-inf.mpg.de. https://vcai.mpi-inf.mpg.de/projects/StyleRig/. Accessed 13 Nov 2023.

Tomlinson, K. (2021). *Vickie doesn't exist: deepfakes in your LinkedIn.* Ampere Industrial Security. [online]. https://www.amperesec.com/newsarchive/deepfakes-in-your-linkedin. Accessed 12 Nov 2023.

United Nations. (2020). *The impact of digital technologies.* [online]. www.un.org. https://www.un.org/en/un75/impact-digital-technologies

Vaccari, C., & Chadwick, A. (2020, May 28). Analysis | 'deepfakes' are here. These deceptive videos erode trust in all news media. *Washington Pos*t. [online]. https://www.washingtonpost.com/politics/2020/05/28/deepfakes-are-here-these-deceptive-videos-erode-trust-all-news-media/

Van den Oord, A., Kalchbrenner, N., Espeholt, L., Vinyals, O., & Graves, A. (2016). Conditional image generation with pixelcnn decoders. *Advances in Neural Information Processing Systems, 29.*

van der Sloot, B., & Wagensveld, Y. (2022). Deepfakes: Regulatory challenges for the synthetic society. *Computer Law & Security Review, 46*, 105716. [online]. https://doi.org/10.1016/j.clsr.2022.105716

Verdoliva, L. (2020). Media forensics and deepfakes: An overview. *IEEE Journal of Selected Topics in Signal Processing, 14*(5), 910–932. https://doi.org/10.1109/jstsp.2020.3002101

Westerlund, M. (2019). The emergence of deepfake technology: A review. *Technology Innovation Management Review, 9*(11), 39–52. [online]. https://doi.org/10.22215/timreview/1282

Whittaker, L., Mulcahy, R., Letheren, K., Kietzmann, J., & Russell-Bennett, R. (2023). Mapping the deepfake landscape for innovation: A multidisciplinary systematic review and future research agenda. *Technovation, 125*, 102784. [online]. https://doi.org/10.1016/j.technovation.2023.102784

Willsher, K., & Holmes, O. (2019). Conmen made €8m by impersonating French minister—Israeli police. *The Guardian.* [online]. https://www.theguardian.com/world/2019/mar/28/conmen-made-8m-by-impersonating-french-minister-israeli-police. Accessed 12 Nov 2023.

Zeitchik, S. (2021, October 15). Every movie and TV show could soon be dubbed into any language you want. *Washington Post.* [online]. https://www.washingtonpost.com/business/2021/10/14/dubbing-film-ai-innovations/

Securing Cyberspace: Addressing the Dark Web and Cybercrime Underreporting

Reza Montasari and Bethany Hopcraft

Abstract The rise of cybercrime presents significant obstacles to effective digital policing, particularly within the Dark Web where anonymity fuels illicit activities. Law enforcement agencies (LEAs) struggle to identify and prosecute criminals in this clandestine realm. As a result, they might resort to ethically dubious methods to identify and apprehend cybercriminals, leading to ethical dilemmas. Moreover, underreporting of cybercrime, especially among older adults with limited technical knowledge, exacerbates these challenges. This chapter critically analyses the challenges posed by the Dark Web to digital policing and cybercrime mitigation. It offers recommendations for LEAs by emphasising the implementation of a Decentralised Identity Management System (DIMS) to enhance user privacy and security in the Dark Web. Additionally, insights from exploratory research on older adults' preferences suggest strategies for effective cybercrime awareness and prevention. By addressing these issues, this research contributes to bolstering cyber safety and law enforcement efforts in the digital realm.

Keywords Cybercrime · Digital policing · Dark Web · Anonymity · Underreporting · Older adults · Decentralised Identity Management

1 Introduction

Cybercrime has emerged as a pervasive and escalating societal concern, necessitating the attention of law enforcement and digital policing to effectively address this burgeoning issue (Curtis & Oxburgh, 2022; McIntyre & Montasari, 2022; Montasari, 2024a). According to reported statistics, in 2019 alone, one million computer misuse crimes occurred in households in England and Wales, emphasising the gravity of the situation (Curtis & Oxburgh, 2022, as cited in National Crime Agency, 2020). The Dark Web plays a pivotal role in facilitating these crimes by enabling perpetrators to

R. Montasari (✉) · B. Hopcraft
Department of Criminology, Sociology and Social Policy, School of Social Sciences, Swansea University, Singleton Park, Swansea, UK
e-mail: Reza.Montasari@Swansea.ac.uk

H. Jahankhani et al. (eds.), *Space Law Principles and Sustainable Measures*, Space Law and Policy, https://doi.org/10.1007/978-3-031-64045-2_9

maintain anonymity and conceal their illicit activities (Davies, 2020). Nevertheless, the reported figures do not adequately reflect the actual prevalence of cybercrime due to underreporting by victims (Jones, 2022). Consequently, this pervasive underreporting poses a significant challenge in addressing cybercriminal activities (Bidgoli et al., 2019).

To effectively combat cybercrime, it is imperative to gain a comprehensive understanding of the issues involved and explore strategies to enhance the efficacy of digital policing (Montasari et al., 2023; Montasari, 2023a, 2023b). Therefore, this research seeks to identify specific measures to improve the effectiveness of digital policing. To this end, the objectives of this study are twofold: first, to examine the challenges posed by the Dark Web and propose potential strategies to address these obstacles; second, to investigate the phenomenon of underreporting in cybercrime and suggest viable solutions to mitigate this issue.

The first objective will be achieved by examining the mechanisms through which the Dark Web provides anonymity, offering insights into the complexities of investigating cybercriminal activities. Additionally, an exploration of the existing policing strategies employed for the Dark Web and their limitations will be undertaken. The proposed recommendation seeks to enhance user privacy and security on the Dark Web while implementing safeguarding measures to bolster policing efforts. The second objective aims to investigate the primary method of reporting cybercrime to the police and identify potential flaws in the process. Moreover, an examination of the reasons behind the underreporting of cybercrime, with a specific focus on the lack of knowledge and understanding of cyberspace, will be conducted. Subsequently, the chapter proposes a strategy to increase reporting rates by raising awareness and providing preventive measures to vulnerable populations. Noteworthy contributions of this chapter include innovative approaches to adapt the Dark Web ecosystem and insights into the preferences of older adults regarding cybercrime education and support.

2 Background

2.1 Cybercrime and Digital Policing

To address the challenges in digitally policing cybercrime, a foundational understanding of the concept of cyberspace and its constituent elements is essential (McIntyre & Montasari, 2022). Cybercrime is broadly defined as "any crime that is facilitated or committed using a computer, network, or hardware device." (Gordon & Ford, 2006, p. 14). Cybercrime can be categorised into three classifications as represented in Fig. 1 (Wall, 2007). Crimes in the device pertain to the presence of illegal content within a machine, while crimes using the device involve offenses committed through a machine, particularly in online environments. Crimes against the device encompass incidents where the device or system is compromised, directly targeting

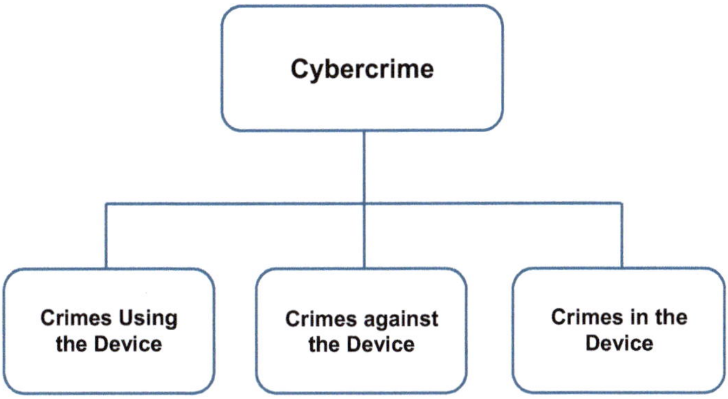

Fig. 1 The three classification of cybercrime adapted from Wall (2007)

the confidentiality, integrity, and availability of systems and data, namely the CIA triad (Wall, 2007).

Digital policing assumes the responsibility of addressing cybercrime (Baraz & Montasari, 2023; McIntyre & Montasari, 2022; Montasari & Boon, 2023). In order to carry out digital policing in a meaningful manner, it is imperative that victims report the crime to appropriate authorities, such as Action Fraud. Crime reports play a crucial role in providing law enforcement with information about the victims and the modus operandi of the cybercrime (Bele et al., 2014). Despite the significance of reporting, empirical data from 2018 reveals a disconcerting statistic, indicating that only 13% of cyber-related frauds were reported to the police or the National Fraud Reporting Center for England and Wales during that year (Office of National Statistics [ONS], 2022; Dodge & Burruss, 2019). This underreporting phenomenon is attributed to the complex nature of cybercrime, leading to a substantial number of victims choosing not to report incidents (Hargittai et al., 2019).

2.2 *Dark Web*

World Wide Web can be categorised into three tiers: the Surface Web, the Deep Web, and the Dark Web (Chertoff & Simon, 2015; Montasari, 2024a). The Surface Web is easily accessible to the public and is subject to global government monitoring. Constituting 90% of internet traffic, the Deep Web is encrypted with passwords and remains unindexed by search engines. It serves as a repository for blocked sites and restricted-access networks, including government files (Chertoff, 2017). Within the Deep Web lies the Dark Web, also known as the 'Darknet' (Chertoff & Simon, 2015). Comprising intentionally concealed, unindexed content accessible solely through specialised software, the Dark Web constitutes a mere 0.01% of Internet content, housing objectionable material due to its anonymity and lack of surveillance (Chertoff, 2017; Davis & Arrigo, 2021).

Originally established by the U.S. Naval Research Laboratory to facilitate secure online communication between military units and field agents, the Dark Web requires specialised anonymity software for access. The Onion Router Project, commonly known as Tor, is a frequently used software for this purpose (Dutta et al., 2021; Moggridge & Montasari, 2022; Montasari & Boon, 2023). Although Tor is legal to use for accessing the Dark Web, engaging in certain activities on the platform is illegal. Notably, 96.6% of Tor network users leverage it for surveillance-free browsing. This highlights the Tor network's role in supporting ethical practices that promote democracy, freedom, and privacy (Jardine, 2015). Examples include providing a platform for whistleblowers (Schein & Trautman, 2019).

However, despite its legitimate applications, the Dark Web has become an ideal space for cybercriminal activities. Unlike the Surface Web, policing the Dark Web poses considerable challenges (Bamsey & Montasari, 2022; McIntyre & Montasari, 2022; Montasari, 2023a, 2023b), as identifying and locating offenders is virtually impossible (Dodge & Burruss, 2019). Consequently, this complexity has led to the acceptance of unconventional investigative techniques in digital investigations to safeguard societal well-being (Chertoff & Jardine, 2021).

3 Challenges

Law enforcement encounters numerous challenges in their efforts to combat cybercriminal activities. This section analyses two primary obstacles: the influence of the Dark Web and the underreporting of cybercrime, exploring their ramifications on the efficacy of digital policing.

3.1 The Impact of the Dark Web

The policing of cybercrimes presents a significantly more complex challenge compared to traditional physical crimes, primarily attributable to the anonymity afforded by the Dark Web (William et al., 2022). The Dark Web ensures anonymity through the use of a Virtual Private Browser (VPN) and Tor (Davies, 2020). A VPN, functioning as an online service, allows users to connect to various networks, and when used in conjunction with Tor, it conceals the user's Internet Protocol (IP) address and encrypts their data. The Tor network further complicates tracking by routing user requests through multiple global machines, encrypting and decrypting user identification in the process, thereby rendering the IP address extremely difficult to trace (Davies, 2020). This robust anonymity mechanism impedes government agencies and other Dark Web users from accessing the user's information or location, highlighting the significant challenges in tracking cybercriminal activities (Kovalchuk et al., 2021).

Originally designed to provide a platform for anonymous browsing and hosting confidential information exchanges, the Dark Web has been leveraged unethically for nefarious purposes (Moore & Rid, 2016). For instance, although it serves as a refuge for whistleblowers to expose wrongdoing while safeguarding their identity (Schein & Trautman, 2019), it has also become an ideal haven for cybercriminals to perpetrate and conceal their illicit activities. The Dark Web facilitates the creation of accounts for communication and trade among criminals through anonymised network connections (Dodge & Burruss, 2019). The spectrum of crimes committed in this space spans from simple electronic theft to heinous offenses such as human trafficking, the distribution of child pornography, and the provision of assassination services (Kovalchuk et al., 2021). Thus, these complex challenges posed by the Dark Web significantly complicate the task of digital policing in combating cybercrime (Montasari, 2022, 2024b; Montasari & Jahankhani, 2021; Staley & Montasari, 2021).

Therefore, conventional policing tools and methods, proven effective in addressing traditional crimes, prove inadequate when utilised in an attempt to investigate cybercrimes (Greco & Montinaro, 2021). Unlike tangible evidence prevalent in physical crimes, the dematerialised environment of the Dark Web necessitates novel investigative approaches (Greco & Montinaro, 2021). In practice, this has often resulted in the reliance on ethically contentious techniques (Chertoff & Jardine, 2021). One such investigative method employed on the Dark Web is the use of honeypot traps. These are sites set up by LEAs to display illicit content, enticing criminals to access and download it, thereby enabling the acquisition of users' IP addresses (Haasz, 2016).

The success of using a honeypot trap method is exemplified by the takedown of the darknet child abuse site, Playpen, in 2015 (Chertoff & Jardine, 2021). The FBI, in a controversial move, allowed the site to operate for an additional 13 days, infecting the digital machines of visitors with malware. While over 8,000 machines were identified, leading to hundreds of arrests, subsequent research has questioned the ethical implications of running the site for an extended period (McIntyre & Montasari, 2022). The unintentional consequence was the further victimisation of children, as over 10,000 users gained access and distributed the illicit content. Although the FBI obtained a warrant to seize the server belonging to the site's creator, they were not authorised to monitor the site (Chertoff & Jardine, 2021).

The FBI justified the operation by emphasising the need to identify individuals beyond the site creator, yet concerns persist about the agency's failure to prevent the continued uploading of harmful content during the monitoring period (Cox, 2017, as cited by Chertoff & Jardine, 2021). The Playpen investigation serves as a stark illustration of the profound challenges posed by the Dark Web's anonymity, compelling law enforcement to resort to ethically questionable methods to apprehend cybercriminals.

3.2 Underreporting of Cybercrime

The general public is well-informed about the need to contact the police in the event of traditional physical crimes; however, the same level of awareness is less ingrained in relation to cybercrime (Bidgoli & Grossklags, 2016). Despite an annual occurrence of over 4.6 million cases of fraud, with 53% transpiring online (ONS, 2022 as cited in Jones, 2022), the reporting of cybercrime remains limited, with only 413,945 reports received by Action Fraud in 2020 (Action Fraud, 2022). Addressing this underreporting is crucial as reporting is a pivotal step towards finding solutions to cybercrimes (Bidgoli & Grossklags, 2016).

The significance of reporting cybercrime is underscored by several justifications (Bidgoli et al., 2019). Firstly, reports contribute valuable information about the frequency of cybercrimes. Secondly, they play a role in exposing the various forms of cybercrimes, including those causing emotional, psychological, or financial harm. Thirdly, reported information can prompt the development of techniques to prevent cybercrime and support victims (Bidgoli et al., 2019). Additionally, reports aid LEAs in determining how to address cybercrime and allocate resources (Bele et al., 2014).

To facilitate reporting, Action Fraud serves as the National reporting Centre for fraud and cybercrime in the UK, established in 2013 as the primary contact point for victims. They broadly define cybercrime as an 'online crime', simplifying the identification of relevant activities for individuals and organisations affected (Action Fraud, 2023). Action Fraud provides a 24-h web chat for assistance and guidance, offering a similar service for charities and organisations online. However, reporting via phone is limited to weekdays between 8 am and 8 pm. Despite emphasising the importance of reporting incidents within the first 72 h, individuals are advised to call emergency services at 999 in case of immediate danger or risk of harm (Action Fraud, 2023). In addition, this timeframe might not be suitable for distressed victims, a common experience leading to long-lasting emotional and psychological effects such as post-traumatic stress disorder (PTSD) (Button et al., 2020).

Nevertheless, the 24-h webchat is designed to address this issue. After a report is filed, the information is forwarded to the City of London's Police Department of National Fraud Intelligence Bureau (NFIB). The NFIB evaluates the information to determine whether there exists sufficient grounds to pursue the case and subsequently forwards it to the local police force where the crime occurred. However, the transnational nature of cybercrimes poses challenges, enabling criminals to victimise individuals globally (Greco & Montinaro, 2021) and leading to judicial complications due to conflicting laws and protocols (Arnell & Faturoti, 2022). Therefore, addressing underreporting and streamlining reporting processes are crucial for enhancing the effectiveness of combating cybercrime on a global scale.

Various reasons have been identified why individuals might refrain from reporting instances of cybercrime, with a predominant factor being a lack of awareness that they have fallen victim to such activities (Karagiannopoulos et al., 2021). This lack of awareness can be attributed in part to insufficient knowledge and understanding about cybercrimes and cybersecurity (Bidgoli & Grossklags, 2016). A notable vulnerability exists among 60-year-olds and above in the UK, who are more likely to become victims of cybercrime due to a general unfamiliarity with cyberspace, as highlighted by Carlson (2019).

According to an Age UK survey, approximately 500,000 individuals aged 65 and above in the UK experienced victimisation to cybercrime in 2015 (Karagiannopoulos et al., 2021). Despite this, these vulnerable populations are compelled to navigate the internet due to the increasing push for essential services to be accessed online, including banking and utility bill payments (Karagiannopoulos et al., 2021). However, challenges arise in the literature exploring older adults' vulnerability to cybercrime due to variations in the age definitions used across studies. The European Union (2022) defines older adults as those aged 50 and above, while the Age UK poll designates older adults as those aged 65 and above. This discrepancy creates a challenge for digital policing and LEAs seeking to protect vulnerable groups from cybercrime, as the research lacks consensus on the age range considered vulnerable.

However, contrary to common assumptions, the underreporting of cybercrime is not solely attributed to older adults. Emotional factors, such as embarrassment and a decline in self-esteem resulting from experiences such as romance scams, contribute significantly to the reluctance of individuals to report cybercrimes to authorities (Nurse, 2018). On the other hand, some victims express scepticism about the police's capacity to effectively address cybercrime, leading to a belief that law enforcement might be ill-equipped to handle their cases (Curtis & Oxburgh, 2022; Huff et al., 2010). These complex reasons for underreporting underscore the complexities involved in addressing cybercrime and necessitate targeted strategies to encourage reporting and enhance public trust in law enforcement capabilities.

4 Recommendations

4.1 Decentralised Identity Management Systems

Certain scholars advocate for enhancing the digital investigative powers of the police as a strategy to counter cybercriminal activities on the Dark Web (Davies, 2020). Proposing the development of policies aimed at increasing the investigative capabilities of digital policing, these suggestions encounter challenges related to potential violations of human rights, exemplified by the Playpen case (Chertoff & Jardine, 2021). Faced with these challenges, government and LEAs might contemplate the prohibition of the Dark Web entirely (Davis & Arrigo, 2021). The primary recommendation identified in this study is the implementation of a Decentralised Identity Management (IdM) System for the Dark Web, specifically advocating for the adoption of the uPORT Open-Source Identity Management System (Naik & Jenkins, 2017).

Decentralised IdM Systems afford users the ability to manage their online identities, granting them the choice to disclose or withhold personal information during online interactions (Maram et al., 2021). In the evolution of identity management systems, the Centralised IdM system, designed for closed environments, relied on central authorities, leading to the development of the Federated IdM System. This system enables single sign-in across multiple organisations. However, users were still limited in controlling their identity. The subsequent evolution resulted in the development of the Decentralised IdM Systems, also known as the Self-Sovereign IdM system (Ruff, 2018), offering users global control over their own identity and data (Naik & Jenkins, 2017). Implementing such a system could address the challenges posed by the anonymity of the Dark Web on digital policing.

One notable implementation of a Decentralised IdM System is the uPORT Open-Source Identity Management System on the Ethereum blockchain. This system has successfully allowed users to exchange information privately while benefiting from the security features of blockchain technology (Naik & Jenkins, 2023). The blockchain serves as a decentralised ledger for transactions and data in cryptocurrencies, ensuring security and anonymity by eliminating the need for third-party organisations in managing these transactions (Yli-Huumo et al., 2016). The uPORT system enables users to conduct both on-chain and off-chain interactions, facilitating simple blockchain transactions, such as buying shares, and private statements with other users (Naik & Jenkins, 2023). Despite concerns raised about the potential risks associated with users having too much control over their identity (Dib & Toumi, 2020), the uPORT system is designed to provide enhanced security and control over identity keys (Braendgaard, 2017). Alternatively, other systems such as the CanDID system proposed by Maram et al. (2021) could also be effective in providing practical, user-friendly decentralised identity solutions, empowering users to control their online identity and data.

However, the integration of IdM systems into the Dark Web faces overarching challenges, due to the complex nature of this part of cyberspace (William et al., 2022). IdM systems, designed to be created and maintained by the user (Maram et al., 2021), would inherently clash with the criminal activities prevalent on the Dark Web. Criminals utilising the Dark Web are unlikely to willingly adopt an IdM system, as it would risk exposing their illicit activities. Therefore, for an IdM system to be viable in the Dark Web, it would necessitate comprehensive integration across the entire space, requiring more stringent access controls than those offered by Tor. The Dark Web operates as a decentralised and complex network, making the implementation of a uniform IdM system a challenging task. While the conceptual premise of an IdM system could potentially address existing issues of anonymity on the Dark Web, the practical application is hindered by the complexity and clandestine nature of this online space (Montasari, 2024c; William et al., 2022).

Therefore, considering these Decentralised IdM systems could offer a potential avenue to balance privacy and security concerns in the context of the Dark Web. Additionally, the integration of IdM systems into the Dark Web could theoretically enhance digital policing by addressing anonymity concerns. However, the feasibility of such implementation is hampered by the complex and decentralised structure of the Dark Web. This emphasises the need for comprehensive understanding and innovative approaches to navigate the challenges associated with policing cybercrime in this unique and complex online environment.

4.2 Addressing the Underreporting of Cybercrime

The underreporting of cybercrime has a detrimental impact on the efficacy of digital policing efforts (Bidgoli et al., 2019). Individuals lacking knowledge and understanding of cyberspace are often unaware of being victimised. This in turn leads to a reluctance to report cybercrime incidents (Bidgoli & Grossklags, 2016). Therefore, to enhance the effectiveness of digital policing, a recommendation is to promote cybercrime prevention and awareness support, educating individuals on what cybercrime entails, how to prevent it, and the importance of reporting (Bele et al., 2014). This support aims to empower individuals to efficiently report cybercrimes, aiding digital police officers in providing assistance to a greater number of victims and potentially apprehending more criminals (Bele et al., 2014).

In the UK, efforts are underway to incorporate cybercrime and cybersecurity education, with some universities offering hands-on laboratory activities simulating real-world cyber-related problems. While studies indicate that educating young people has improved their knowledge of cybercrime (European Union, 2022), it is acknowledged that focusing solely on this demographic might not adequately address the underreporting issue (Karagiannopoulos et al., 2021). Therefore, a more comprehensive approach is needed to include a wider range of age groups.

Notably, older adults are identified as a vulnerable population susceptible to cybercrime victimisation. This emphasises the need for tailored cybercrime awareness and prevention support for this demographic (Bele et al., 2014). Current cybercrime support for older adults tends to be generic and may not adequately address their specific needs (Karagiannopoulos et al., 2021). A study conducted by the Portsmouth Cybercrime Awareness clinic revealed significant findings regarding older adults' attitudes and experiences with cybercrime and cyber-awareness support (Karagiannopoulos et al., 2021). The study highlighted high levels of concern among participants, linking victimisation to a lack of knowledge and feelings of loneliness. Despite this, over half of the participants had not received age-appropriate cybercrime awareness or prevention support. Those who did receive support found it often irrelevant and lacking specificity for their age group (Karagiannopoulos et al., 2021).

Importantly, the study revealed a general lack of awareness regarding where to report cybercrimes and seek support, indicating a need for improved awareness campaigns targeting older adults (Karagiannopoulos et al., 2021). This suggests that older adults might not be aware of resources such as Action Fraud, the National reporting centre that offers support (Action Fraud, 2023). Therefore, addressing these gaps in awareness and tailoring support services to the unique needs of older adults are crucial for mitigating underreporting and fostering a safer digital environment for this demographic.

Despite being particularly vulnerable to cybercrime (Carlson, 2019), there is a notable gap in knowledge concerning the preferences of older adults for receiving cybercrime support (Nicholson et al., 2019). According to the research conducted by Karagiannopoulos et al. (2021), participants expressed a preference for incorporating routine cybercrime education. The consensus among the participants was that educational sessions could prove effective, and potential locations identified included cafes, supermarkets, nursing homes, and libraries. Additionally, participants recommended online sessions for adults who are housebound. Some individuals proposed one-on-one meetings with IT professionals or the organisation of awareness talks and events. A key emphasis was placed on the importance of resources being presented in a simple and easily understandable manner, avoiding complex language (Karagiannopoulos et al., 2021).

Therefore, to increase the effectiveness of digital policing, it is essential to promote cybercrime prevention and awareness support with a specific focus on older adults and their preferences. This approach should aim to provide a range of methods for older adults to receive support tailored to their technological experience, physical abilities, and social demands (Karagiannopoulos et al., 2021). While underreporting is influenced by various factors beyond knowledge of cybersecurity and cybercrime (Nurse, 2018), dedicating resources to directly address this issue could be highly beneficial. Adopting this approach would enhance the digital policing of cyberspace, as cybercrime support for older adults can contribute to reducing loneliness, preventing victimisation, enabling individuals to identify whether they have been victimised, and providing the necessary knowledge and resources to report incidents appropriately. This, in turn, allows digital policing to operate more efficiently and provides LEAs with better insights into the targets of cybercrime, aiding in determining the necessary staffing levels to combat these digital offenses.

5 Conclusion

In conclusion, policing cybercrime, especially within the anonymity of the Dark Web, presents significant challenges, compounded by the issue of underreporting. The Dark Web facilitates horrific crimes, and LEAs are forced to employ unconventional and sometimes ethically questionable practices to catch offenders (McIntyre & Montasari, 2022). The suggestion of implementing Decentralised IdM systems (Naik & Jenkins, 2017) aligns with the need for a more sustainable approach to policing the Dark Web. However, the complex nature of the Dark Web poses a challenging obstacle to various strategies aimed at law enforcement (Chertoff & Simon, 2015; Montasari et al., 2023). The underreporting of cybercrime is a complex issue influenced by factors such as a lack of awareness and understanding (Bele et al., 2014), emotional barriers (Nurse, 2018), and negative perceptions of law enforcement (Curtis & Oxburgh, 2022).

Older adults, in particular, face challenges due to their lower levels of technical knowledge, making them more susceptible to cybercrimes. Recommendations to promote cybercrime prevention and awareness support, especially tailored for older adults, not only are beneficial for digital policing but also contribute to societal well-being. These initiatives can empower legitimate Dark Web users with privacy and protection while providing ongoing assistance to older adults, boosting their confidence in navigating the digital landscape and reducing their vulnerability to cybercrimes and social isolation (Karagiannopoulos et al., 2021; Naik & Jenkins, 2023). Overall, addressing these challenges requires a comprehensive and multi-layered approach that combines technological solutions, educational initiatives, and support services.

References

Action Fraud. (2022). *Annual assessment fraud crime trends 2020/21*. Action Fraud. https://data.actionfraud.police.uk/cms/wp-content/uploads/2021/07/2020-21-Annual-Assessment-Fraud-Crime-Trends.pdf

Action Fraud. (2023). *Reporting fraud and cyber crime*. Action Fraud UK; Metropolitan Police. https://www.actionfraud.police.uk/reporting-fraud-and-cyber-crime

Arnell, P., & Faturoti, B. (2022). The prosecution of cybercrime—Why transnational and extraterritorial jurisdiction should be resisted. *International Review of Law, Computers & Technology, 37*(1), 1–23. https://doi.org/10.1080/13600869.2022.2061888

Bamsey, O., & Montasari, R. (2022). A critical analysis of the Dark Web challenges to digital policing. In H. Jahankhani, S. Kendzierskyj, R. Montasari, & N. Chelvachandran (Eds.), *Social media analytics, strategies and governance* (pp. 192–202). CRC Press.

Baraz, A., & Montasari, R. (2023). Law enforcement and the policing of cyberspace. In R. Montasari, V. Carpenter, & A. J. Masys (Eds.), *Digital transformation in policing: The promise, perils and solutions* (pp. 59–83). Springer. https://doi.org/10.1007/978-3-031-09691-4_4

Bele, J. L., Dimc, M., Rozman, D., & Jemec, A. S. (2014). Raising awareness of cybercrime—The use of education as a means of prevention and protection. In *International Conference Mobile Learning*.

Bidgoli, M., & Grossklags, J. (2016). End user cybercrime reporting: What we know and what we can do to improve it. In *IEEE International Conference on Cybercrime and Computer Forensic* (pp. 1–9). https://doi.org/10.1109/ICCCF.2016.7740424

Bidgoli, M., Knijnenburg, B. P., Grossklags, J., & Wardman, B. (2019). *Report now. Report effectively. Conceptualizing the industry practice for cybercrime reporting*. IEEE Xplore. https://doi.org/10.1109/eCrime47957.2019.9037577

Braendgaard, P. (2017). *What is uPORT identity? Medium*.

Button, M., Sugiura, L., Blackburn, D., Shepard, D. W. J., Wang, V., & Kapend, R. (2020). *Victims of computer misuse: Main findings*. Home Office. https://www.gov.uk/government/organisations/home-office

Carlson, E. (2019). Phishing for elderly victims: As the elderly migrate to the internet fraudulent schemes targeting them follow by Eric L. Carlson. *The Elder Law Journal*. https://theelderlawjournal.com/2019/01/28/carlson/

Chertoff, M. (2017). A public policy perspective of the Dark Web. *Journal of Cyber Policy, 2*(1), 26–38.

Chertoff, M., & Jardine, E. (2021). *Policing the Dark Web legal challenges in the 2015 playpen case* (pp. 1–14). CIGI Papers.

Chertoff, M., & Simon, T. (2015). *The impact of the Dark Web on internet governance and cyber security*. Global Commission on Internet Governance.

Curtis, J., & Oxburgh, G. (2022). Understanding cybercrime in "real world" policing and law enforcement. *The Police Journal: Theory, Practice and Principles, 96*(4), 573–592. https://doi.org/10.1177/0032258x221107584

Davies, G. (2020). Shining a light on policing of the Dark Web: An analysis of UK investigatory powers. *The Journal of Criminal Law, 84*(5), 407–426. Sage Journals. https://doi.org/10.1177/0022018320952557

Davis, S., & Arrigo, B. (2021). The Dark Web and anonymizing technologies: Legal pitfalls, ethical prospects, and policy directions from radical criminology. *Crime, Law and Social Change, 76*, 367–386. https://doi.org/10.1007/s10611-021-09972-z

Dib, O., & Toumi, K. (2020). Decentralized identity systems: Architecture, challenges, solutions and future directions. *Annals of Emerging Technologies in Computing, 4*(5), 19–40. https://doi.org/10.33166/aetic.2020.05.002

Dodge, C., & Burruss, G. (2019). Policing cybercrime: Responding to the growing problem and considering future solutions. In *The human factor of cybercrime* (pp. 339–358). Routledge.

Dutta, N., Jadav, N., Tanwar, S., Sarma, H. K. D., & Pricop, E. (2021). TOR—The Onion Router. *Studies in Computational Intelligence*, 37–55. https://doi.org/10.1007/978-981-16-6597-4_3

European Union. (2022). *Special Eurobarometer 464b: Europeans' attitudes towards security*. Data.europa.eu. https://data.europa.eu/data/datasets/s1569_87_4_464b_eng?locale=en

Gordon, S., & Ford, R. (2006). On the definition and classification of cybercrime. *Journal in Computer Virology, 2*(1), 13–20. https://doi.org/10.1007/s11416-006-0015-z

Greco, G., & Montinaro, N. (2021). The phenomenon of cybercrime: From the transnational connotation to the need for globalisation of justice. *European Journal of Social Sciences Studies, 6*(1), 1–10. https://doi.org/10.46827/ejsss.v6i1.956

Haasz, A. (2016). Underneath it all: Policing international child pornography on the Dark Web. *Syracuse Journal of International Law & Commerce, 43*(2), 353–380. https://web.s.ebscohost.com/ehost/pdfviewer/pdfviewer?vid=0&sid=898510c7-ffcb-4a47-9948-25189c316777%40redis

Hargittai, E., Piper, A. M., & Morris, M. R. (2019). From internet access to internet skills: Digital inequality among older adults. *Universal Access in the Information Society*, 881–900.

Huff, R., Desilets, C., & Kane, J. (2010). *National public survey on white collar crime*. National White Collar Crime Center.

Jardine, E. (2015). *The Dark Web dilemma: Tor, anonymity and online policing* (Global Commission on Internet Governance Paper Series, No. 21). https://doi.org/10.1007/s11277-020-07143-2

Jones, P. (2022). *Nature of crime: Fraud and computer misuse*. Office of National Statistics. https://www.ons.gov.uk/peoplepopulationandcommunity/crimeandjustice/datasets/natureofcrimefraudandcomputermisuse

Karagiannopoulos, V., Kirby, A., Oftadeh-Moghadam, S., & Sugiura, L. (2021). Cybercrime awareness and victimisation in individuals over 60 years: A Portsmouth case study. *Computer Law and Security Review, 43*.

Kovalchuk, O., Masonkova, M., & Banakh, S. (2021). The Dark Web worldwide 2020: Anonymous vs safety. In *International Conference on Advanced Computer Information Technologies* (pp. 526–530). https://doi.org/10.1109/acit52158.2021.9548578

Maram, D., Malvai, H., Zhang, F., Jean-Louis, N., Frolov, A., Kell, T., Lobban, T., Moy, C., Juels, A., & Miller, A. (2021, May 1). *CanDID: Can-do Decentralized identity with legacy compatibility, sybil-resistance, and accountability*. IEEE Xplore. https://doi.org/10.1109/SP40001.2021.00038

McIntyre, M. W., & Montasari, R. (2022). The Dark Web and digital policing. *Artificial Intelligence and National Security*, 193–203. https://doi.org/10.1007/978-3-031-06709-9_10

Montasari, R. (Ed.). (2022). *Artificial Intelligence and national security*. Springer. https://doi.org/10.1007/978-3-031-06709-9

Montasari, R. (Ed.). (2023a). *Applications for Artificial Intelligence and digital forensics in national security*. Springer. https://doi.org/10.1007/978-3-031-40118-3

Montasari, R. (2023b). *Countering cyberterrorism: The confluence of Artificial Intelligence, cyber forensics and digital policing in US and UK national cybersecurity*. Springer. https://doi.org/10.1007/978-3-031-21920-7

Montasari, R. (2024a). *Cyberspace, cyberterrorism and the international security in the fourth industrial revolution: Threats, assessment and responses*. Springer.

Montasari, R. (2024b). Assessing the effectiveness of UK counter-terrorism strategies and alternative approaches. In R. Montasari (Ed.), *Cyberspace, cyberterrorism and the international security in the fourth industrial revolution: Threats, assessment and responses* (pp. 27–50). Springer.

Montasari, R. (2024c). Exploring the current landscape of cyberterrorism: Insights, strategies, and the impact of COVID-19. In R. Montasari (Ed.), *Cyberspace, cyberterrorism and the international security in the fourth industrial revolution: Threats, assessment and responses* (pp. 65–90). Springer.

Montasari, R., & Boon, A. (2023). An analysis of the Dark Web challenges to digital policing. In H. Jahankhani (Ed.), *Cybersecurity in the age of smart societies* (pp. 371–383). Springer. https://doi.org/10.1007/978-3-031-20160-8_19

Montasari, R., Carpenter, V., & Masys, A.J. (Eds.). (2023). *Digital transformation in policing: The promise, perils and solutions*. Springer. https://doi.org/10.1007/978-3-031-09691-4

Moggridge, E., & Montasari, R. (2022). A critical analysis of the Dark Web challenges to digital policing. In R. Montasari (Ed.), *Artificial Intelligence and national security* (pp. 157–167). Springer, Cham. https://doi.org/10.1007/978-3-031-06709-9_8

Montasari, R., & Jahankhani, H. (Eds.). (2021). *Artificial Intelligence in cyber security: Impact and implications: Security challenges, technical and ethical issues, forensic investigative challenges*. Springer. https://doi.org/10.1007/978-3-030-88040-8

Moore, D., & Rid, T. (2016). Cryptopolitik and the Darknet. *Survival, 58*(1), 7–38. https://doi.org/10.1080/00396338.2016.1142085

Naik, N., & Jenkins, P. (2017). Securing digital identities in the cloud by selecting an appropriate federate identity management from SAML. In *11th International conference on Research Challenges in Information Science (RCIS)* (pp. 163–174). IEEE.

Naik, N., & Jenkins, P. (2023). *uPORT open-source identity management system: An assessment of self-sovereign identity and user-centric data platform built on blockchain* (pp. 1–6). IEEE Xplore.

Nicholson, J., Coventry, L., & Briggs, P. (2019). "If it's important it will be a headline": Cybersecurity information seeking in older adults. In *Proceedings of the 2019 CHI conference on human factors in computing systems*. https://doi.org/10.1145/3290605.3300579

Nurse, J. R. C. (2018). Cybercrime and you: How criminals attack and the human factors that they seek to exploit. In *The Oxford handbook of cyberpsychology* (pp. 662–690). https://doi.org/10.1093/oxfordhb/9780198812746.013.35

Office of National Statistics. (2022). *Nature of fraud and computer misuse in England and Wales*. Office of National Statistics. https://www.ons.gov.uk/peoplepopulationandcommunity/crimeandjustice/articles/natureoffraudandcomputermisuseinenglandandwales/yearendingmarch2022#trends-in-computer-misuse

Ruff, T. (2018). *The three models of digital identity relationships*. Medium. https://medium.com/evernym/

Schein, D. D., & Trautman, L. J. (2019). The Dark Web and employer liability. *SSRN Electronic Journal*. https://doi.org/10.2139/ssrn.3251479

Staley, B., & Montasari, R. (2021). A survey of challenges posed by the Dark Web. In R. Montasari & H. Jahankhani (Eds.), *Artificial Intelligence in cyber security: Impact and implications* (pp. 203–213). Springer, Cham. https://doi.org/10.1007/978-3-030-88040-8_8

Wall, D. S. (2007). Policing cybercrimes: Situating the public police in networks of security within cyberspace. *Police Practice and Research, 8*(2), 183–205. https://doi.org/10.1080/15614260701377729

William, P. E., Jawale, M. A., Pawar, A. B., Priyanka, N., & Bibave, R. R. (2022). Systematic approach for detection and assessment of Dark Web threat evolution. In *IGI Global EBooks* (pp. 230–256). https://doi.org/10.4018/978-1-6684-6444-1.ch013

Yli-Huumo, J., Ko, D., Choi, S., Park, S., & Smolander, K. (2016). Where is current research on blockchain technology?—A systematic review. *PLOS One, 11*(10). https://doi.org/10.1371/journal.pone.0163477

MIX
Papier aus verantwortungsvollen Quellen
Paper from responsible sources
FSC® C105338

If you have any concerns about our products,
you can contact us on
ProductSafety@springernature.com

In case Publisher is established outside the EU,
the EU authorized representative is:
Springer Nature Customer Service Center GmbH
Europaplatz 3, 69115 Heidelberg, Germany

Printed by Libri Plureos GmbH
in Hamburg, Germany